Nuclear Power Plants

Edited by **Matt Fulcher**

CLANRYE INTERNATIONAL

New Jersey

Published by Clanrye International,
55 Van Reypen Street,
Jersey City, NJ 07306, USA
www.clanryeinternational.com

Nuclear Power Plants
Edited by Matt Fulcher

© 2015 Clanrye International

International Standard Book Number: 978-1-63240-389-6 (Hardback)

Printed in the United States of America.

Contents

Preface

Every book is initially just a concept; it takes months of research and hard work to give it the final shape in which the readers receive it. In its early stages, this book also went through rigorous reviewing. The notable contributions made by experts from across the globe were first molded into patterned chapters and then arranged in a sensibly sequential manner to bring out the best results.

Nuclear power plant is described as a power plant in which nuclear energy is converted into heat for use in producing steam. The main challenge faced globally is gathering enough energy for the increasing population and controlling the carbon emission caused by the usage of fossil fuels. To overcome this, nuclear energy can be used as an alternative power source. This book includes research work and technical experience from different power plants and research institutions around the world. It discusses various topics like nuclear systems protection, design and modeling of critical parameters in nuclear power plants, thermal-hydraulic analysis, nuclear waste management and safety assessment. The book consists of significant information which will be of interest to scientists and those in technical areas of nuclear power plants.

It has been my immense pleasure to be a part of this project and to contribute my years of learning in such a meaningful form. I would like to take this opportunity to thank all the people who have been associated with the completion of this book at any step.

Editor

Power System Protection Design for NPP

Chang-Hsing Lee and Shi-Lin Chen

Additional information is available at the end of the chapter

1. Introduction

One of the key purposes of NPP power system protection is to ensure that NPP's local power demand (such as cooling pumps, control systems, etc.) are met under all circumstances even during faulted periods. To achieve this goal, NPP power system protection must ensure that it can supply these local loads using either (1) power from the grid (via the transmission connection, which in most time, however, are used for exporting the excess power generated by the NPP after supplying its local loads) or (2) power from local generations such as diesel generators, batteries, etc. at all times and under all circumstances.

On the first power source (grid power), many NPPs worldwide have been built along the seashore for cooling water availability reasons. Overhead transmission lines are thus built in the vicinity of the seashore to transport the large amount of power generated from the NPP to the grid economically. As these overhead lines are exposed to salt contamination, flashover will occur when contamination becomes excessive. In the event of flashover, which is equivalent to a line-to-ground fault, the plant's protection system will need to initiate a series of switching operation to redirect the large power output from the NPP to a backup route in order to avoid reactor emergency shut-down. However, such switching has the adverse effect of causing undesirable transient overvoltages to propagate in the plant's local power grid [1-4]. Dealing with the frequent switching actions of these overhead lines while mitigating their adverse effects thus becomes the first challenge of designing NPP power system protection.

Once the NPP loses its connection to the grid, it will need to rely on the local generation to continue supplying its local loads. Most NPP use multiple "independent" sources as backup power. However, unless NPP's local power grid is properly configured and its protection system properly designed, these "independent" sources can all fail at the same time as manifested in Taipower's 18 March, 2001 Level 2 event ("318 Event") [5].

In the following sections, we will examine Taipower's "318 Event" in detail to demonstrate the various possibilities that could lead to NPP plant blackout. Moreover, as these possibilities are not mutually exclusive, we will use this example to illustrate how multiple or cascaded problem can present further challenges to the overall NPP power system protection design. Recommended preventive measures are then summarized in the final section of this chapter.

2. Taipower "318 Event"

2.1. System configuration

Figure 1 shows the configuration of Taipower 3rd nuclear power plant. The NPP has two 951 MW generators which are connected to the local 345kV gas-insulated substation (GIS) in one-and-half breaker configuration as shown in Fig. 1. The NPP is then connected to the power grid via four 345 kV overhead power lines (Darpen 1, 2 and Lunchi Sea/Mountain) to the Darpen and Lunchi 345kV EHV (Extra High Voltage) substation and two 161kV overhead lines (Kengting and Fengkang) to Kenging and Fenkang 161kV HV (High Voltage) substation.

It is important to note that there are three 13.8 kV buses (in the middle) and four 4.16 kV buses (at the bottom) for plant utility. Among these buses, the two 4.16 kV buses in the middle are responsible for feeding the safety-critical equipment such as cooling pumps and are designated as "essential buses". (The 2nd 4.16kV bus from the left where "DGA (Diesel Generator A)" is connected is designated as "Essential Bus A". The one next to it, where "DGB" is connected to, is "Essential Bus B".)

Another notable but subtle feature of this configuration is the use of 3-phase gas-insulated line (GIL) design with the 3 phases enclosed in a single duct of approximately 340 meters for the connection of the generation units and auxiliary systems, (located at the foot of a hill) to the 345kV GIS (on the top of the hill) due to topography feature of the location. This feature has implication on the generation and propagation of switching transients which will be explained in later sections.

2.2. Event sequence

On 18 March, 2001, a Level 2 event occurred at Taipower's 3rd NPP, and the whole plant went into blackout from 00:45 to 02:58. The event started at 00:45 when EHV CB3510 (see Fig. 1, highlighted in red) was closed to energize the then-offline 345 kV/13.8 kV/4.16 kV start-up transformer (X01). Upon CB3510 closure, medium voltage (MV) CB#17 on "Essential Bus A" exploded damaging not only CB#17 but also CB#15. CB#15 formed a permanent ground fault keeping "Essential Bus A" at ground potential thus Essential Bus A became useless. CB #3 and #5 on "Essential Bus B" was then opened hoping DGB will start and supply power to those critical loads. However, DGB failed to start and the whole plant went into blackout. The only hope remained at that time was DG5 (on the far right in Fig. 1) which, however, needs to be started locally and manually. After 2 hour since the first problem occurred, DG5 was finally started and started to supply power to the critical loads via "Essential Bus B".

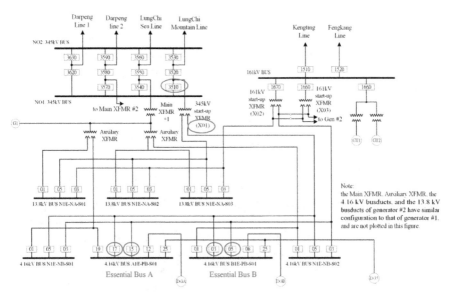

Figure 1. System Configuration of the NPP

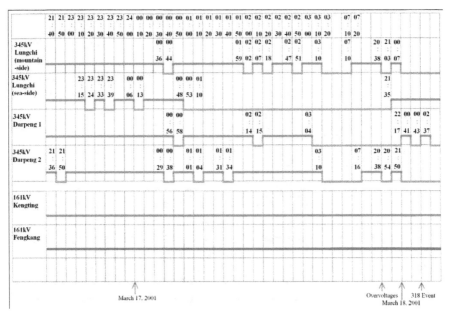

Figure 2. 345kV and 161kV Overhead Line Switching Event Log

Figure 2 shows the switching event log of the four 345kV and two 161kV lines connecting to the NPP. After reviewing the event log, it was found that CB#17 on Essential Bus A broke down when one GIS switching operation was occurring. The event log also showed that there were 37 EHV switching operations during the 48-hour period prior to the event due to salt-fog influence in the plant area. Because of the unstable offsite power, the GIS switched between different offsite power to acquire the stable power sources.

Figure 3 shows the transient recording of overvoltages for both the 345kV bus and one MV (medium voltage) bus at 20:38 in March, 2001. At t0, the flashover on the 345kV line occurred leading to its subsequent tripping at t1. The tripping took place on the remote end of the 345kV line thus overvoltage can still be observed at the NPP between t1 and t2 due to "motor-generating effects" to be explained in the following section of this Chapter. The overvoltage on the 345kV line eventually caused flashover from Phases A and B to ground pulling down the line voltage and all motors on the 4.16kV bus were tripped by their respective under voltage relays at t3. At t4, the flashover from Phases A and B to ground was cleared and the "motor-generating effects" start to build up the voltages again with the two remaining motors on the 13.8kV bus.

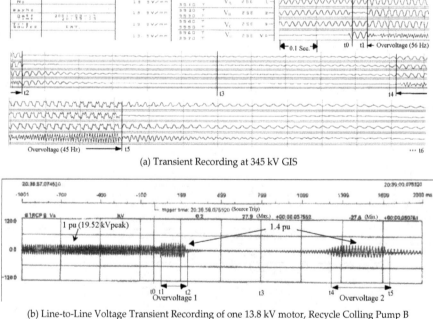

(a) Transient Recording at 345 kV GIS

(b) Line-to-Line Voltage Transient Recording of one 13.8 kV motor, Recycle Colling Pump B

Figure 3. Overvoltages at 20:38 in March 17, 2001

2.3. Electrical stress in plant power system

2.3.1. Line conductor overvoltages due to over-excitation and nonlinear resonance [6,7]

The transient recorder in Fig. 3(a) recorded 2 abnormal overvoltages (at 56 Hz and 45 Hz, respectively) after the last 345 kV-transmission line connecting to the NPP was tripped on the remote end which turned the NPP into an electrical island. As will be explained in the next section, the 1st overvoltage was caused by the over-excitation of the motors (e.g. recycle water pump) in the plant who, with terminal voltages supported by large line capacitance, now operated as induction generator after loss of external power.

The 2nd overvoltage is caused by a different mechanism. After a few cycles the low voltage relays tripped many of the plant motors leaving only 2 biggest motor (now operating as induction generator) still connected and were supported by a comparatively much larger capacitance leading to not only over-excitation but also magnetic saturation of both the motors and transformers. This created a condition very close to ferroresonance resulting in even bigger overvoltage.

2.3.2. Neutral voltage transfer

It can be seen from Figure 3 that overvoltage were observed not only on the line conductors of phase A, B, and C but also on the neutral. As will be explained in the next section, neutral voltage transfer can occure through 2 different mechanisms: electromagnetic and capacitive transfer.

In the presence of transformer core saturation, 3rd harmonic neutral voltage will be present on the windings through electromagnetic transfer as long as the neutrals of the respective windings are not grounded. In the presence of neutral voltage on any of the transformer windings, the stray capacitance among the windings and earth will result in capacitive neutral voltage transfer.

2.3.3. Switching surges on both EHV and MV systems

From Fig. 2, it can be seen that there were 37 switching operations during the 48-hour period prior to the breakdown. This unusually high number of switching operation can create lots of switching surges (with magnitude of around 7 times the rated line-to-ground peak voltage in the medium voltage system) which, when propagating through the NPP local power network, can degrade the insulation level or even cause breakdown of CB's in the local power network [8,9].

It should be noted that while there were 37 switching operations on the EHV side, none of the switching surges were captured by the transient recorder in the 345kV GIS in Fig. 3(a) due to insufficient bandwidth of the transient recorder. In a follow-up field test [9] after the event, it was found that such switching often causes switching surges of around 7 times the rated line-to-ground peak voltage!

3. Stress mechanism and modeling

It can be seen from the above that Taipower's 3rd NPP was under sigificant and multiple stresses before and during the Level 2 event. This section explains the mechanisms working behind these stresses and provide basic principles how to model them.

3.1. Line conductor overvoltages due to over-excitation and nonlinear resonance [6,7]

Figure 3 shows that on Phases A, B, and C there were two overvoltages observed where the second overvoltage was slightly higher than the first. Causes of these 2 overvoltages are detailed as following.

3.1.1. First overvoltage (56Hz) – Over excitation

Figure 4 shows the 2 essential condition for induction motor generating effect: large capacitance and continuous rotating motor. When an induction motor lost its external voltage source, the flywheel with large inertia will keep the motor rotating and the capacitance of transmission line will provide the necessary voltage support for the induction motor to act as a generator. The magnitization curve of the motor and the amount of capacitance will jointly determine the overall motor generating effect as shown in Fig. 5. If the capacitance is too small to provide enough magnetizing current (curve C0 in Fig. 5), the terminal voltage of motor will decay exponentially and the generating effect will not sustain. However, if the capacitance is large enough, the motor generating effect will sustain and the   terminal voltage is determined by the intersection of the capacitance and magnetizing curve such as (V1, C1) and (V2, C2) in Fig. 5.

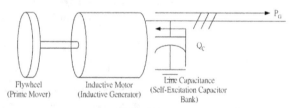

Figure 4. Equivalent Circuit of Motor-Generating Effect.

At "t1" in Fig. 3(a), the last 345kV transmission line connecting to the NPP was tripped on remote end due to a flashover on the line turning the NPP into an electrical island. As the local end of the 345kV line did not trip, a motor generating condition equivalent to Fig. 4 was formed with the 127kM transmission line providing sufficient capacitance to support the voltage of the various motors in the NPP. As can be seen in Fig. 3(b), the terminal voltage is increased to 1.4 p.u. and the overall resultant frequency is 56 Hz.

During this first overvoltage period, the terminal voltage of motor was about 1.4 p.u. (Fig. 3(b)) but the line voltage was about 1.29 times the rated line-to-ground peak voltage (Fig.

3(a)). This implied that the power transformers have saturated. As a result, a lot of harmonics were produced and the zero sequence components of them would be integrated into the neutral voltage resulting in unexpected high neutral voltage. This period ended at "t2" in Fig. 3(a) when the flashover grounded both phase A and B.

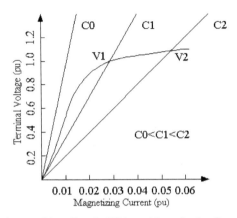

Figure 5. Relationships between Motor Terminal Voltage, Magnetization Curve, and External Capacitance

Table 1 shows the harmonic contents of B phase voltage between t1 and t2 in Fig. 3(a). The even order harmonics and DC component could be treated as the slight magnetic bias caused by asymmetric fault. At this stage, there was no ferromagnetic resonance in the island system.

Order	DC	1	2	3	4	5	6	7	8
%	9.3	100	7.8	8.0	3.6	13.5	1.6	6.1	1.2

Table 1. Voltage Harmonic Contents of Phase B between t1 and t2

3.1.2. Second overvoltage (45 Hz) - Nonlinear resonance

Figure 6 shows the four essential conditions for a ferroresonance to occur: voltage source, capacitance, nonlinear inductance (ferromagnetic and saturable), and low losses. The R in the RLC resonant circuit in Fig. 6 is very large due to the "low losses" condition and can often be ignored. The nonlinear inductance L is the magnetizing curve of the motors and transformers in the system and the capacitance is provided by the transmission line.

At "t3" in Fig. 3(a), all the motors on the 4.16 kV system were tripped by undervoltage relay. Between t3 and t4, the flashover grounding of phases A and B were cleared and the motor generating effects mentioned above picked up again gradually re-establishing the line voltage. At "t4" in Fig. 3(a), most motors in 13.8 kV system were also tripped by undervoltage relay with the exception of two largest ones. With the capacitance provided by the transmission line now need only to support the terminal voltage of 2 motors, we would expect the terminal voltages to be higher than those during the first overvoltage stage according to Fig. 5. However, due to deep saturation of the motors and transformers, the overvoltage magnitude in Fig. 3(b) during the 2nd overvoltage is only slightly higher than the previous stage. This can be further seen from the fact that at the beginning of "t4" in Fig. 3(a), there were no overvoltage and no distortion of waveforms. As line voltage increased, the harmonics increased and after a few cycles the amplitude of voltage remained but voltage waveform distorted dramatically. Figure 7 shows the waveform at 4 cycle prior to t5 with its Fourier components summarized in Table 2 [10].

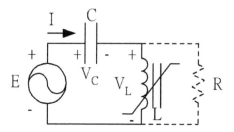

Figure 6. Equivalent RLC Circuit

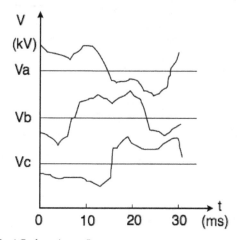

Figure 7. Zoom-in of The 4 Cycles prior to t5

It can be seen from Table 2 that the voltage of fundamental frequency was about 1.5 times the rated line-to-ground peak voltage, which is slightly larger than the previous overvoltage. The large DC and even-order harmonics indicate the deep saturation of the start-up transformer. In Table 2 the total of 3rd harmonics is 554.7 kV_{peak} (Note: The 3rd harmonics are in-phase therefore can be added up directly.) Comparing this figure with the neutral voltage of 626 kV_{peak} in Fig. 3(a), this indicates that 3rd harmonics is the main source of neutral voltage during the second overvoltage period.

order	Phase A		Phase B		Phase C	
	V_{peak}	%ofund.	V_{peak}	%ofund.	V_{peak}	%ofund.
0	175.068	0.429	48.279	0.113	71.93	0.017
1	408.273	1.0	427.991	1.0	413.726	1.0
2	108.386	0.265	47.287	0.11	39.829	0.096
3	179.679	0.44	179.734	0.42	195.357	0.472
4	41.29	0.101	28.313	0.066	24.702	0.06
5	28.13	0.069	30.829	0.072	65.626	0.159
6	13.036	0.044	10.364	0.024	22.34	0.054
7	12.562	0.031	22.126	0.052	37.579	0.091
8	15.84	0.039	28.219	0.066	7.708	0.019
9	14.315	0.035	8.207	0.019	39.509	0.095
10	9.428	0.023	16.289	0.038	6.203	0.015
11	18.092	0.044	14.962	0.035	27.434	0.066
12	18.805	0.046	27.966	0.065	25.323	0.061
13	15.387	0.038	11.466	0.027	22.233	0.054
14	15.717	0.038	10.323	0.027	24.685	0.06
15	13.897	0.037	24.991	0.058	13.998	0.034
16	1.992	0	36.961	0.086	12.173	0.029

Table 2. Fourier Analysis of Fig. 7

As the inductances in the systems are now deeply saturated, there is a possibility that ferroresonance can occur. (Note: Ferroresonance is nonlinear resonances in power system where the voltage and current may change from normal steady state to another steady state with large harmonic distortion.) The phenomenon can be best understood from a circuit perspective using Figure 6 as example. In Fig. 6, the total equivalent impedance of the circuit is (jX_L - jX_C). When the inductance is saturated and current further increases, it will drive the inductor into deeper saturation where the inductor impedance jX_L will reduce when current further increases. A critical point will be reached at Point B in Figure 8 when (jX_L - jX_C) becomes zero. Any current increase beyond Point B will cause the total impedance change from a positive value to a negative value causing resonance effects near this operating point[7].

Based on analysis of all available data, it is believed that the 2nd overvoltage from t4 onward is on the boundary to be ferroresonance therefore the 2nd overvoltage is caused by a combination of motor-generating effect and nonlinear resonance.

3.2. Neutral voltage transfer [10]

It can be seen from Figure 3 that overvoltage can be observed not only on the line conductor but also on the neutral conductor as well. In order to understand this phenomenon we need to look at Fig. 9 where the equivalent circuit of a transformer is shown including its stray capacitances.

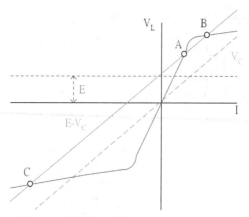

Figure 8. Ferroresonance Phenomenon Explanation

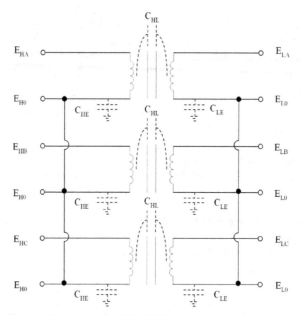

Figure 9. Voltage Transfer Diagram of 345 kV/4.16 kV Transformer

3.2.1. Transformer modeling

In Fig. 9, C_{HE} and C_{LE} depict the stray capacitance between high voltage (HV) winding to ground, and low voltage (LV) winding to ground, respectively, while C_{HL} depicts the stray capacitance between HV and LV windings. Typical stray capacitances for the 345/13.8/4.16kV power transformer are shown in Table 3.

In the presence of transformer core saturation, 3rd harmonic neutral voltage will be present on the windings through electromagnetic transfer as long as the neutrals of the respective windings are not grounded. Once the neutral voltage is established on any side of the neutrals, the stray capacitance provides a further path for it to transfer to other neutrals according to Equation (1)

$$E_{L0} / E_{H0} = C_{HL} / (C_{HL} + C_{LE}) \qquad (1)$$

where E_{H0} is the neutral voltage at HV side, and E_{L0} is the neutral voltage at LV side.

Item	$C_{345/Earth}$	$C_{13.8/Earth}$	$C_{4.16/Earth}$	$C_{345/13.8}$	$C_{13.8/4.16}$	$C_{345/4.16}$
Capacitance	4.48 nF	13.76 nF	21.92 nF	4.3 nF	214.86 pF	8.96 nF

Table 3. Stray Capacitance of the 345kV/13.8kV/4.16kV Power Transformer

3.2.2. Neutral voltage transfer

It can be seen from Fig. 3(a) that, during 1st overvoltage the neutral voltage gradually roses to 200 kVrms while during 2nd overvoltage the neutral voltage rose to 626 kVpeak (Note: the voltage waveform became very non-sinusoidal during 2nd overvoltage, we thus use peak value instead of rms value). The source of both overvoltages in the neutral was due to motor and transformer saturation resulting in 3rd harmonic voltages at the neutral however during the 2nd overvoltage the waveform is much more distorted with higher harmonic content.

As indicated by Fig. 3(a), the neutral on 345kV side does not appear to have been effective grounded possibly due to grounding failure. The result is that very high neutral voltage was established on the 345kV neutral and if the 4.16kV neutral was not grounded it will see a neutral voltage (through capacitive neutral transfer) of

$$200kVrms * \frac{8.96nF}{8.96nF + 21.92nF} = 58.03kVrms$$

$$626kVpeak * \frac{8.96nF}{8.96nF + 21.92nF} = 181.54kVpeak$$

during 1st and 2nd overvoltages, respectively. To better appreciate various grounding combination's effect on the neutral voltage transfer, 3 simulations were conducted assuming grounding conditions as per Table 4 and their results are summarized in Table 5.

Item	345kV side	13.8kV side	4.16kV side
Case 1	Ground (direct)	Ground (8 Ω)	Ground (2.4 Ω)
Case 2	Non-Ground	Non-Ground	Non-Ground
Case 3	Non-Ground	Ground (8 Ω)	Ground (2.4 Ω)

Table 4. Grounding Condition for Simulating Capacitive Transfer

Table 5 shows that as the neutral voltages transferred to the 4.16kV bus can be as high as 13 times the phase-to-ground peak voltage which can pose significant threat to CB#17 as well as other CB's. However, if the neutral systems were properly configured, the risk can be minimized greatly.

	345kV side Phase Voltage	345kV side Neutral Voltage	4.16kV side Neutral Voltage
Case 1	457.55 kV$_{peak}$	0 kV$_{peak}$	0 kV$_{peak}$
Case 2	450.6 kV$_{peak}$	275.88 kV$_{peak}$	42.29 kV$_{peak}$
Case 3	448.93 kV$_{peak}$	44.58 kV$_{peak}$	0.002 kV$_{peak}$

(a) 1st overvoltage

	345kV side Phase Voltage	345kV side Neutral Voltage	4.16kV side Neutral Voltage
Case 1	428.83 kV$_{peak}$	0 kV$_{peak}$	0 kV$_{peak}$
Case 2	418.7 kV$_{peak}$	263.43 kV$_{peak}$	41.08 kV$_{peak}$
Case 3	420.94 kV$_{peak}$	46.25 kV$_{peak}$	0.002 kV$_{peak}$

(b) 2nd overvoltage

Table 5. Capacitive Transfer Simulation Result

3.3. Switching surges and Very Fast Transient Overvoltage (VFTO)

Switching operations are the most prominent phenomenon in the "318 Event". During the 48 hours prior to the Level 2 event, there were 37 switching operations and each could cause switching surges. Switching surges caused by GIS switching is characterized by its nanosecond wavefront and is commonly referred to as Very Fast Transient Overvoltage (VFTO) [11].

VFTO is the phenomenon of transient overvoltage generated during switching operation characterized by very short rise-time of 4 to 100 ns and has been covered by various literatures [3,12-22]. The phenomenon is particularly significant during Disconnect Switch (DS) operation due to multiple-restriking in the DS due to lack of arc-suppressing chamber.

3.3.1. Field measurement

In the past, VFTO was not considered to be possible to transfer from EHV through power transformer to medium voltage (MV) system [12-14]. However, in light of the "318 Event", a field test was conducted in Taipower 3rd NPP during plant overhaul by switching the DS of EHV GIS and measure the voltage on "Essential Bus A".

Field test result [9] shows that after switching the DS of EHV GIS, multiple 25 kV-level restrikes (approximately 7 times the rated line-to-ground peak voltage) were measured on the 4.16kV bus indicating VFTO can be transferred from the EHV side to MV side. It also indicates that the maximum peak voltages measured on the 4.16kV bus occur neither on the first strike nor on the last strike, and this behaviour is quite different with that in EHV system. The measurement results are shown in Fig. 10.

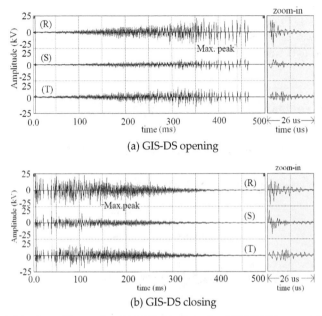

Figure 10. Switching Surge Measured on 4.16kV Bus by Operating EHV GIS Disconnect-Switch. (Note: The bandwidth of the measurement system was 2MS/s, the highest achievable in 2003)

3.3.2. VFTO simulation

To further appreciate VFTO transfer mechanism, numerical simulation model was built [23]. To validate this simulation model, the field test condition for Fig. 10 was reconstructed and the simulation result is shown in Fig. 11. It can be seen from Fig. 11 that the waveform envelope are consistent with measurement for both DS opening and closing and that the maximum VFTO on the essential bus occurred neither at first nor at last strike.

We then change the DS operation angle for each 5° intervals to simulate different closing/opening condition and Table 6 and 7 summarizes the maximum EHV inter-contact breakdown voltage vs. maximum MV VFTO. The following can be observed from Table 6 and 7:

1. The VFTO transferred to the essential bus A can be as high as 28.77 kV, which is about 8.47 times the rated line-to-ground peak voltage.
2. For all simulations, the restrike that causes the maximum VFTO on "Essential Bus A" does not necessarily coincide with the one that caused the max inter-contact breakdown voltage on EHV side.
3. Among the 36 simulations for DS opening, the simulation that produces the highest inter-contact breakdown voltage on EHV side is not the same as the one that produces

the maximum VFTO on "Essential Bus A". This is also true for DS closing. E.g., Case #28 (δ_{oper} = 135°) of DS opening produces the highest VFTO in MV system (28.77 kV) while it was Case #18 (δ_{oper}=85°) that produces the highest inter-contact breakdown voltage on EHV side (354.2 kV).

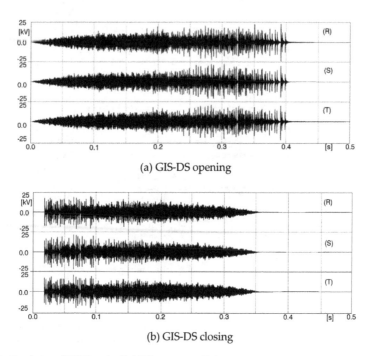

(a) GIS-DS opening

(b) GIS-DS closing

Figure 11. Simulation of VFTO at the Field Measurement Point

Item	Among the Multiple Restrikes				Total Num. of Restrikes on EHV Side per Φ
	Max. Inter-contact Breakdown Voltage		Max. VFTO at 4.16 kV		
	Mag. (kV)	Seq. Num.	Mag. (kV)	Seq. Num.	
Max. Inter-contact Breakdown Voltage Case # 18 among 36	354.2	467	23.76	444	468
Max. VFTO at 4.16 kV Bus Case #28 among 36	323.6	447	28.77	432	447

Table 6. Max. Inter-contact Breakdown Voltages vs. Max. VFTO in MV for DS Opening

Item	Among the Multiple Restrikes				Total Num. of Restrikes on EHV Side per Φ
	Max. Inter-contact Breakdown Voltage		Max. VFTO at 4.16 kV		
	Mag. (kV)	Seq. Num.	Mag. (kV)	Seq. Num.	
Max. Inter-contact Breakdown Voltage Case #14 among 36	284.8	1	23.27	12	460
Max. VFTO at 4.16 kV Bus Case # 23 among 36	278.0	1	26.41	26	455

Table 7. Max. Inter-contact Breakdown Voltages vs. Max. VFTO in MV for DS Closing

3.3.3. Characteristic of VFTO transferring to MV system

3.3.3.1. Capacitive coupling of high-turn-ratio transformer

VFTO and the oscillation voltages V_{OSC} (voltages created by preceding restriking that can be superimposed to the following restrike) on the EHV side can be transferred to MV system through the start-up power transformer via capacitive coupling. The transfer ratio is mainly dependent on transformer's EHV-to-MV interwinding capacitance, transformer's MV winding-to-enclosure capacitance, and the bus-to-ground capacitance of MV system [23]. From both our measurement and simulation result, it was observed that the V_{OSC}, which is of several tens kV in the EHV GIS, could still be of several kV in the MV system, and this will be superimposed to the VFTO coupled from the EHV side causing up to 7 ~ 8.47 times the rated line-to-ground peak voltage on MV side.

3.3.3.2. Superposition of oscillations initiated by a prior strike on top of subsequent restrikes

Figure 12(a) shows two consecutive restrikes from a multiple-restrike simulation and Fig. 12(b) shows its counterpart single-strike simulation. It can be seen from Fig. 12(a) that the V_{OSC} initiated by the first restrike is superimposed to the second restrike resulting in a higher peak voltage (10.72 kV vs. the single strike one of 9.88kV).

3.3.3.3. Maximum VFTO transferred to MV for DS closing vs. DS opening

During DS opening the contact distance becomes wider and wider leading to longer intervals between two consecutive restrikes while that during DS closing is the opposite. As a result, there is a higher probability of superposition of V_{OSC} to subsequent restrike during DS closing (thus higher VFTO) than opening.

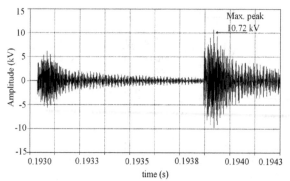

(a) Two consecutive restrikes taken out of multiple-restrike simulation of DS opening

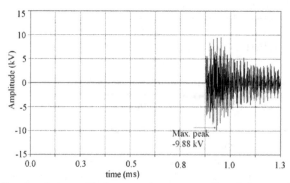

(b) Single-strike simulation with the same inter-contact breakdown voltage as in (a)

Figure 12. Oscillation Voltage (Vosc) Initiated by a Strike or Restrike Can Be Superimposed to Subsequent Restrike Voltages.

4. Lesson learned and important aspects of NPP power system protection design

Events like the "318 Event" were seldom caused by one single reason. It can be seen from the above discussion that Taipower 3rd NPP was under multiple stresses before the event and there were mutiple mechanisms for the generation, amplification, and transfering of overvoltages which, combined with the operation practices and equipment history, eventually led to the explosion of CB#17 and total blackout of the NPP. Below are the key lessons learned from this event and their recommended preventive measure.

4.1. Bus configuration and fault area isolation

The "318 Event" was essentially triggered by a single equipment failure but leading to a complete blackout of the power plant. There are 2 key lessons learned from this event: (1)

Explosion of CB#17 took down the adjacent CB#15 as well. (2) "Independent sources" are not always independent due to improper bus configuration.

For various reason such as space requirement, ease of maintenance, etc, switchgear panels are usually installed in the same room side by side. If this cannot be changed, during the risk evaluation process one must consider the N-1 condition being loss of "one group of equipment" instead of "one equipment" unless sufficient separation are provided between the equipments.

The "independence" of power sources need then be examined closely. If multiple sources or multiple buses can be taken down by a single failure such as permenant fault to ground, etc, they cannot be considered as independent sources and more backup needs to be added.

It should be noted that during the "318 Event", after the explosion of CB#17 the plant utility room was filled with smoke which makes the manual starting of other diesel generators extremely difficult. Not only were equipments under significant stress but also the human operators. It is thus recommended that the feasiblity of starting backup sources under utility room smoke condition be checked and that any manual operation required during this stage be as simple and straightforward as possible with proper interlock to reduce the chance of human error which may further escalate the event.

4.2. Nonlinear resonance prevention

Among all the scenarios considered in this Chapter, nonlinear resonance is the most difficult one to be detected. In view of the potential hazard it could cause, precautionary measure must be taken to prevent it from initiating.

The first step is to prevent motor-generating effect from ever occuring (thus removing the key source of initiation.) As explained above, the essential conditions of motor-generating effect are (1) rotating motor with large inertia, (2) large capacitor bank in an isolation system to support the terminal voltage. Since a rotating motor with large inertia can not be stop immediately, the focus is to remove the capacitive support. In the case of Taipower 3rd NPP, the capacitive support came from the long transmission line who were tripped only on the remote end. It is recommended that Direct Transfer Trip (DTT) function be implemented for transmission line protection to greatly reduce the risk of motor generating effect.

The second step is to ensure effective grounding of transformer neutrals as designed. Due to the objective of minimizing short circuit current, the neutral groundings in NPP are usually multi-configured: arrestor grounded under normal condition and direct grounding when in islanding operation. The switching from one grounding scheme to another often requires manual operation and this increases the risks of leaving the islanded system ungrounded as well as nonlinear resonance of power. Proper interlock or checking mechanism should be implemented to ensure proper grounding as designed at all times.

4.3. Neutral voltage transfer

Neutral voltage transfer can occur via either electromagnetic or capacitive transfer. Based on simulation result the risk can be significantly reduced with proper grounding of the neutral. This, however, must be carefully implemented in order not to increase the short circuit current in the NPP.

Again, any manual operation during event would introduce extra risks therefore should be designed to be as simple and straightforward as possible with proper interlock or checking system.

4.4. VFTO transferring to MV system

According to field measurement and numerical simulation, the VFTO transferring to MV system is usually underestimated by literatures. As demonstrated by both the field measurement and simulation result, peak voltage of VFTO in MV system could be as high as 8.47 times the rated line-to-ground peak voltage with an average 466 times restrike during DS operating [1,23]. Though the peak VFTO voltages transferred to the MV side are usually still within the basic impulse insulation level (BIL) tolerances of the equipment, this does not mean that repeatedly striking the equipment with 8.47 times the rated line-to-ground peak voltage would cause no damages to the equipment. In fact, this can accelerate equipment ageing and cause quick degradation of the insulation material and eventually leading to equipment breakdown.

After the "318 Event", a recommendation was made to Taipower No. 3 Nuclear Power Plant in 2003 for the installation of surge absorbers (0.8μF capacitor specially designed for surge absorption installed right close to the start-up transformer for each of the three phases) on the MV side in Fig. 1 [9,23]. The recommendation was adopted by Taipower in March 2005 and a subsequent measurement in March 2006 plus one-year monitoring indicated that there were no further VFTO exceeding rated line-to-ground peak voltage on the MV system.

4.5. Maintenance testing of in-service equipments

The damaged circuit breaker (CB#17) in Taipower 3rd NPP has been put into service for 20 years at the time of event. Maintenance testing history showed that insulation condition of this circuit breaker was good prior to the event however that being the case the circuit breaker should not have exploded when faced by transient voltage no higher than its BIL of 60kV. This shows that the current diagnostic method of insulation degradation (insulation resistance measurement, dielectric power factor measurement) may not be sensitive enough to detect insulation degradation due to ageing or repeated VFTO strikes. It is recommended that the reliability of such tests, including both the tool used, methodology employed, and interpretation of testing results (including monitoring the trend of measurement results) be further improved. For equipment subject to repeated switching surges, a higher standard should be applied.

5. Conclusion

Most NPP's in the world have been designed in such way that their local power loads are provided by "multiple independent sources" to ensure continuous power supply even during faulted periods. However, unless the NPP's local power grid is properly configured and its protection system properly designed, all these "multiple independent sources" can failed at same time as exemplified by Taipower's "318 Event". In view of the many similarities in design and other risk factors for world NPP's, it is of utmost importance that the lessons learned from Taipower's 3rd NPP "318 Event" be properly addressed.

This Chapter examines the Taipower "318 Event" in detail to demonstrate the various possibilities that could leads to NPP blackout. The possibilities investigated include: NPP's location factor, NPP local power grid configuration, cable parameters, switching events, switching surges propagating to MV circuits, ferroresonances, remote tripping, and manual starting difficulties. The lessons learned and proposed countermeasures are summarized in the previous section.

In summary, to ensure the proper design of NPP power protection system, the following 3 considerations must be incorporated:

1. **Check Independence of Equipment and Protection Zone for Various Scenarios:** The "318 Event" was caused by a single CB failure (CB#17) but leading to a complete NPP blackout for over 2 hours. This is mainly due to (1) the breakdown of CB#17 took down CB#15 at the same time due to their physical proximity. (2) The bus configuration cause none of the 2nd, 3rd, or 4th backup power to be available when both CB#15 and #17 both fails and CB#15 created a permanent line to ground fault. (3) The last resort (DG5) was located in a building filled with smoke caused by the CB#17's breakdown making manual starting extremely challenging. (4) The sustained overvoltage in the system could have been avoided should the tripping of the EHV cable be done on both ends of the line instead of just the remote end. All of the above suggest that the independence of equipment and protection zone have failed and needs to be taken into consideration when improving existing or future designs.
2. **Accumulated Equipment Stress Must be Monitored and Considered**: Particularly relevant for NPP's located on the seashore and subject to frequent line switching, equipment stress manifested in the form of reduced insulation level must be subject to more frequent and detailed monitoring. This would include not only absolute value measuring but also trending the measurement so that early signs of equipment weakness can be identified and proper measures be adopted to address it.
3. **Use System Protection Design Approach**: The cause of "motor generating effect", "neutral voltage transfer", "VFTO", and "ferroresonance" occurring during the "318 Event" cannot be addressed one by one and need to be taken into consideration from a system protection perspective. This would include the consideration for using different tripping scheme (such as Direct Transfer Trip on the EHV line) , adding additional protection device (such as installing surge absorbers on the MV bus), as well as reconfiguring the bus connections.

Abbrevious

BIL	Basic Impulse Level
CB	Circuit Breaker
DG	Diesel Generator
DS	Disconnect Switch
DTT	Direct Transfer Trip
EHV	Extra-High-Voltage
GIL	Gas-Insulated Line
GIS	Gas-Insulated Substation
HV	High Voltage
LV	Low Voltage
MV	Medium Voltage
NPP	Nuclear Power Plant
VFTO	Very Fast Transient Overvoltage

Author details

Chang-Hsing Lee
EE Dep. of National Tsing Hua University, Hsinchu, Taiwan

Shi-Lin Chen
EE Dep. of Chung Yuan Christian University, Chung Li, Taiwan

Acknowledgements

The authors wish to thank Professor He Zhao, Advisor to China Electric Power Research Institute, and Dr. Edward Hsi, EE Department, Chung Yuan Christian University, for their valuable comments on this work.

6. References

[1] Lee C. H., Hsu S. C., Hsi P. H., Chen S. L. Transferring of VFTO from EHV to MV System as Observed in Taiwan's No. 3 Nuclear Power Plant. in IEEE Trans. Power Delivery 2011, Vol. 26, No. 2: 1008-1016.

[2] Jakel W., Muller A. B. Switching Transient Levels Relevant to Medium Voltage Switchgear and Associated Instrumentation. in Proc. International Conference and Exhibition on ElectroMagnetic Compatibility, June 12-13, 1999.: 35-40.

[3] Buesch W., Marmonier J., Palmieri G., Chuniaud O., Miesch M. GIS Instrument Transformers: EMC Conformity Tests for a Reliable Operation in an Upgraded Substation. in Proc. Conference on Electric Power Supply Industry, Oct. 23-27, 2000.: 1-7.

[4] Uglesic I., Hutter S., Milardic V., Ivankovic I., and Filiovic-Grcic B. Electromagnetic Disturbances of the Secondary Circuits in Gas Insulated Substation due to Disconnector

Switching. in Proc. International Conference on Power Systems Transients, Sep. 28-Oct. 2, 2003.: 1-6

[5] Atomic Energy Council, The Station Blackout Incident of the Maanshan NPP unit 1: 2001.

[6] Tsao T. P., Ning C. C. Analysis of Ferroresonant Overvoltage at Maanshan Nuclear Power Station in Taiwan. IEEE Transaction on Power Delivery 2006, Vol. 21, No. 2.: 1006-1012.

[7] A. Greenwood, Electrical Transients in Power Systems. John Wiley & Sons, Inc.

[8] Das J. C. Surges transferred through transformers. in Proc. 2002 Annual Pulp and Paper Industry Technical Conference, 17-21 June, 2002.: 139-147

[9] Chen S. L. A study on The Feasibility to Install Surge Absorber at Low Voltage Side of 345 kV and 161 kV Start-up Transformer in The 3rd Nuclear Power Plant. Taiwan Power Company, Research Report, TPC-546-91-2104-10: 2003.

[10] Zhao H., personal communication: 2005.

[11] CIGRE WG 33/13-09 Very Fast Transient Phenomena Associated with Gas Insulated Substations, CIGRE Report: 1988.

[12] Meppeline J., Diederich K. J., Feser K., Pfaff W. R. Very fast transients in GIS. IEEE Trans. Power Delivery 1989, Vo. 4, No. 1.: 223-233.

[13] Fujimoto N., Boggs S. A. Characteristics of GIS Disconnector-Induced Short Risetime Transients Incident on Externally Connected Power System Components. IEEE Trans. Power Delivery 1988, Vol. 3.: 961-970.

[14] Kumar V. V., Thomas J. M., Naidu M. S. Influence of Switching Conditions on The VFTO Magnitudes in a GIS. IEEE Trans. Poer Delivery 2001, Vol. 16, No. 4.: 539-544.

[15] Popov M., Sluis L. van der, Smeets R. P. P., Roldan J. L. Analysis of Very Fast Transients in Layer-Type Transformer Windings. IEEE Trans. Power Delivery 2007, Vol. 22.: 238-247.

[16] Fujita S., Shibuya Y., Ishii M. Influence of VFT on Shell-Type Transformer. IEEE Trans. Power Delivery 2007, Vol. 22, No. 1.: 217-222.

[17] Shibuya Y., Fujita S., Tamaki E. Analysis of Very Fast Transients in Transformers. in Proc. IEE Generation, Transmission and Distribution Conference 2001, Vol. 148, Issue 5.: 377-383.

[18] Ogawa S., Haginomori E., Nishiwaki S., Yoshiida T, Terasaka K. Estimation of Restriking Transient Overvoltage on Disconnecting Switch for GIS. IEEE Trans. Power System 1986, Vol. 1, No. 2.: 95-102.

[19] Rao M. M., Thomas M. J., Singh D. P. Frequency Characteristics of Very Fast Transient Currents in a 245 kV GIS. IEEE Trans. Power Delivery 2005, Vol. 20, No. 4.: 2450-2457.

[20] Smeets R. P. P., Linden W. A. van der, Achterkamp M., Pamstra G. C., Meulemeester E. M. De Disconnector Switching in GIS: Three-Phase Testing and Phenomena. IEEE Trans. Power Delivery 2000, Vol. 15, No. 1.: 122-127.

[21] Christian J., Xie j. Very Fast Transient Oscillation Measurement at Three Gorges Left Bank Hydro Power Plant. in Proc. 2006 International Conference on Power System Technology, 22-26 Oct. 2006.: 1-7.

[22] Ji L. Y., Huang W. H., Zhang Z. Y., Shi W. Analysis and Simulation of Conducted Interference in Three-Phase in One tank GIS. in Proc. 2009 Second Asia-Pacific Conference on Computational Intelligence and Industrial Applications.: 269-299.

[23] Lee C. H. Simulation and Analysis of The Very Fast Transient Overvoltage in Medium Voltage Systems. Ph. D Thesis, National Tsing Hua University, Taiwan: 2011.

Chapter 2

Flow Accelerated Corrosion in Nuclear Power Plants

Wael H. Ahmed

Additional information is available at the end of the chapter

1. Introduction

In general, corrosion is defined as the degradation of a material by means of chemical reactions with the surrounding environment. Several types of corrosion occur in a variety of situations in the nuclear power plants. Some of these types are common such as rusting of steel when located in moist environment, and the other type of corrosion such as flow accelerated corrosion required special treatment due to their impact on the plant safety and reliability. FAC degradation mechanism results in thinning of large areas of piping and fittings that can lead to sudden and sometimes to catastrophic failures, as well as a huge economic loss. FAC is a process caused by the flowing water or wet steam damaging or thinning the protective oxide layer of piping components. The FAC process can be described by two mechanisms: the first mechanism is the soluble iron production (Fe2+) at the oxide/water interface, while the second mechanism is the transfer of the corrosion products to the bulk flow across the diffusion boundary layer. Although the FAC is characterize by a general reduction in the pipe wall thickness for a given piping component, it frequently occurs over a limited area within this component due to the local high area of turbulence. The rate of the metal wall loss due to FAC depends on a complex interaction of several parameters such as material composition, water chemistry, and hydrodynamic.

In general, erosion processes or mechanisms can be categorized as:

i. **Shear stress erosion**: In this category, the surface of a material gets destroyed in single-phase flow at high velocity, by the effects of shear stresses and the variations in the fluid velocity.

ii. **Liquid impact induced erosion**: This form of erosion occurs in two-phase flow by the impingement of liquid droplets entrained in flowing gases or vapours. It can cause damages to power plant condenser tubes, elbows, turbine blades, etc. The wear process can however be avoided by a combination of improvements in: plant design, drying the steam, and the use of more corrosion-resistant steels.

iii. **Flashing-induced erosion:** This form of erosion occurs when spontaneous vapour formation takes place due to sudden pressure changes. Locations where this type of erosion occurs are found in drain and vent lines downstream of control valves.

iv. **Cavitation erosion:** This type of erosion is caused by repeated growth and collapse of bubbles in a flowing fluid as a result of local pressure fluctuations. In the regions of higher pressures downstream, the sudden collapse of gas bubbles results in pressure spikes that may erode the material in their vicinity. These bubbles will however get re-absorbed along the piping system without causing any damage to the downstream straight piping component.

On the other hand, degradation mechanisms involve combined effect of chemical and mechanical processes can be summarized as:

i. **Erosion-corrosion: In this mechanism** a combination of mechanical and chemical material degradation processes take place. The combined effect of the two processes is considered more severe especially in the case when copper alloy heat exchanger tubes are exposed to high velocities.

ii. **Flow-Accelerated Corrosion:** The pipe wall thinning due to this degradation mechanism in carbon steel piping is not due to the mechanical effect only, however, the wall thinning is mainly due to dissolution of normally protective magnetite film that normally forms on the internal pipe wall surface.

1.1. Conditions required for FAC

It has been observed that the following conditions result in FAC degradation in nuclear or fossil power plants:

- Flow conditions: both single- and two-phase flow conditions, with water or water-steam mixture as the flowing liquid, at a temperature of $> 95^\circ$C, and flow velocity is greater than zero.
- Chemistry condition: the flowing liquid should be such that a potential difference exists between the liquid and the carbon steel pipe wall. This difference will be responsible for the dissolution of the protective oxide layer in the flowing stream. A high magnetite solubility and subsequent rapid removal of the magnetite, is facilitated by either demineralised and neutral water or slightly alkalinized water under reducing conditions.
- Material: the pipe material must be carbon steel or low-alloy steel. General practice recommends that for a well-designed system, FAC will be effectively inhibited if steel components are made to contain at least 0.1% chromium.
- Flow Geometry: FAC has been observed to occur downstream of flow-restricting or redirecting geometries like an orifice, sudden contraction, expansion, elbows, reducers, etc.

Failures and accidents due to FAC degradation have been reported at several nuclear power plants around the world since 1981 [1]. However, detailed analysis of the FAC related failures did not start before the severe elbow rapture downstream of a tee occurred at Surry Unit 2 power plant (USA) in 1989, which caused four fatalities and extensive plant damage and resulted in a plant shutdown. In 1999, an extensive steam leakage from the rupture of the shell side of a feed-water heater at the Point Beach power plant (USA) was reported by Yurmanov and Rakhmanov [2] (Figure 1). In 2004, a fatal pipe rupture downstream of an orifice in the condensate system due to FAC occurred in the Mihama nuclear power plant Unit 3 (Japan) [2]. More recently, the pipe failure downstream of a control valve at Iatan fossil power plant in 2007 resulted in two fatalities and a huge capital of plant loss as reported by Moore [3]. Although, a combination of lab research and attempts to correlate lab results with plant experience has been the major efforts made towards the study of FAC mechanism since the 1970's, the lab research only focused on understanding the mechanisms, and correlating experimental results in order to reduce the lab effort and to develop usable forms for the plant engineers. Several well cited correlations used to predict the actual corrosion rates due to FAC in piping systems and incorporated in computer software such as CHECWORKS developed by Electric Power Research Institute (EPRI). Following the abovementioned accidents, most utilities around the world have been following EPRI guideline of improving the flow water chemistry to slow down the rate of damage. Also, in the event of disposition of highly susceptible or damaged areas, utilities have typically taken the following initial steps:

1. Replace individual worn components with other components of same material.
2. Replace individual worn components with other components of FAC resistant material.
3. Replace entire worn lines with other components of same material.
4. Replace entire susceptible lines or the more susceptible portions with other components of FAC resistant material.

The recent review by Ahmed [4] highlighted the significant research conducted on investigating the effect of fluid chemical properties on flow accelerated corrosion (FAC) in nuclear power plants. He concluded that the hydrodynamic effects of single and two-phase flows on FAC have not been thoroughly investigated for many piping components. In order to determine the effect of the proximity between two components on the FAC wear rate, Ahmed [4] has investigated 211 inspection data for 90° carbon steel elbows from several nuclear power plants. The effect of the velocity as well as the distance between the elbows and the upstream components was discussed. Based on the analyzed trends obtained from the inspection data, the author indicated a significant increase in the wear rate of approximately 70% that was identified to be due to the proximity.

Furthermore, the repeated inspections in nuclear power plants have shown that piping components located downstream of flow singularities, such as sudden expansion or contractions, orifices, valves, tees and elbows are most susceptible to FAC damage. This is due to the severe changes in flow direction as well as the development of secondary flow instabilities downstream of these singularities [4]. Moreover, in two-phase flows, the significant phase redistributions downstream of these singularities may aggravate the problem. Therefore, it is important to identify the main flow and geometrical parameters require in characterizing FAC damage downstream of pipe fittings. These parameters are: the geometrical configuration of the components, piping orientation, and the flow turbulence structure which will affect the surface shear stress and mass transfer coefficients.

18" elbow wall thickness decreased from 12.7 to 1.5 mm on feed-water pump inlet at Surry, 1986

Wall Thickness reduced from 10 to 1.5 mm on Feed-water piping at Mihama unit 3, 2004

Failure in a high pressure extraction line at Fort Calhoun in 1997

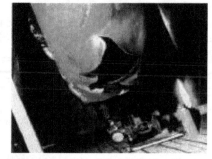

Failure downstream of the LCV in the reheater drain line at Millstone unit 2, 1991.

Failure of 14" heater drain extraction line to high pressure heater at Arkansas unit 2, 1986

Failure of the Feed-water Heater Point Beach Unit 1, 1999

Figure 1. Examples of failures due to FAC worldwide (Yurmanov and Rakhmanov [2])

For single phase flow, the secondary vortices and/or flow separation downstream of pipe fittings considered to be important parameters need to be analyzed and modelled while predicting the highest FAC wear rate location. For example; the secondary flows in elbows induce a pressure drop along the elbow wall that can significantly increase the wall mean and oscillatory shear stresses as discussed by Crawford et al. [5]. Also, orifices and valves promote turbulence close to the wall in the downstream pipe and thus enhance the rate of mass transfer at the wall [5]. These mechanisms have been identified as the governing factors responsible for FAC as explained by Chen et al. [6].

The hydrodynamics parameters controlling FAC in two-phase flows are considered more complex than for single-phase flows due to the complexity of two-phase distribution and the unknown interactions between the gas phase and the liquid [8]. These interactions play a major role in the mass, momentum, and energy transfer between the flow phases as explained by Hassan et al. [9]. Also, the inlet two-phase flow pattern plays an important role in the flow dynamics downstream of the orifices since the phase redistribution downstream depend on the upstream flow regime. For example, bubbles can have significant effects on the turbulent kinetic energy close to the wall, affecting the wall shear stress and pressure. Moreover, Jepson [10] showed that high velocity slugs can cause high turbulence and shear forces at the pipe wall and thus enhance the destruction of the protective inhibitor film.

Computational fluid dynamic (CFD) analysis is used to predict turbulent fluid flow with great accuracy for many applications. However, only the recent advances in computational power have allowed the use of CFD for mass transfer and corrosion studies. This indicated that the accurate prediction of mass transfer near the wall requires resolving the mass transfer boundary layer which may be an order of magnitude smaller than the viscous sub-layer. In order to perform the CFD calculations with good accuracy, fine near-wall grids with correct near-wall turbulence models can therefore provide mass transfer data for the corrosion species. In these cases corrosion is controlled by the mass transfer, relation between the wall mass transfer coefficient and corrosion rate can be derived as explained in details by Keating and Nesic [11]. Furthermore, in formulating the CFD codes, consideration is made to the hydrodynamic parameters affecting the mass transfer rate of the corrosion products to the bulk fluid and consequently the FAC rate. These hydrodynamic parameters are the flow velocity, pipe roughness, piping geometry, and steam quality or void fraction for two-phase flow.

The hydrodynamic effects of the working fluid on FAC have been investigated by many researchers using CFD. Bozzini [12] adopted numerical simulations for investigating wall erosion/corrosion inside a pipe bend for a four-phase flow that comprised of two immiscible liquids, gas and particulate solids. On the other hand, Chang et al. [13] suggested an evaluation scheme to estimate the load carrying capacity of thinned-wall pipes exhibiting FAC. In their study, they employed a steady-state incompressible flow CFD code to determine the pressure distributions as input conditions for a structural finite element analyses in order to calculate local stresses. More recently, Ferng [14] developed an approach that used an erosion/corrosion model and three-dimensional single and two-phase flow models to predict locations of serious FAC in power plant piping systems. Their predictions agreed very well with plant measurements.

2. FAC rate and mass transfer

The FAC process in carbon steel piping is described by three steps. In the first process, metal oxidation occurs at metal/oxide interface in oxygen-free water and explained by the following reactions:

$$Fe + 2H_2O \rightarrow Fe^{2+} + 2OH^- + H_2 \tag{1}$$

$$Fe^{2+} + 2OH^- \leftrightarrow Fe(OH)_2 \tag{2}$$

$$3Fe + 4H_2O \rightarrow Fe_3O_4 + 4H_2 \tag{3}$$

The first process involves the solubility of the ferrous species through the porous oxide layer into the main water flow. This transport across the oxide layer is controlled by the concentration diffusion. The second step is described by the dissolution of magnetite at oxide/water interface as explained by the following reaction:

$$1/3Fe_3O_4 + (2-b)H^+ + 1/3H_2 \leftrightarrow Fe(OH)_b^{(2-b)+} + (4/3-b)H2O) \tag{4}$$

where:

$Fe(OH)_b^{(2-b)+}$ represents the different iron ferrous species $b=(0,1,2,3)$

In the third step (Eq. 4), a diffusion process takes place where the ferrous irons transfer into the bulk flowing water across the diffusion boundary layer. In this process, the species migrated from the metal/oxide interface and the species dissolved at the oxide/water interface diffuse rapidly into the flowing water. In this case, the concentration of ferrous iron in the bulk water is very low compared to the concentration at the oxide/water interface.

It can be noticed that FAC mechanism involves convective mass transfer of the ferrous ions in the water. The convective mass transfer for single phase flow is known to be dependent on the hydrodynamic parameters near the wall interface such as flow velocity, local turbulence, geometry, and surface roughness. In addition, the physical properties of the transported species or the water do not affect the local transport rate in adiabatic flow especially when temperature changes in piping system are negligible. Over a limited length of piping component, FAC rate is considered as direct function of the mass flux of ferrous ions and can be calculated from the convective mass transfer coefficient (MTC) in the flowing water. Then, FAC rate is calculated from the MTC and the difference between the concentration of ferrous ions at the oxide/water interface (C_w) and the concentration of ferrous in the bulk of water (C_b) as:

$$FAC \ rate \ = \ MTC(\ C_w - \ C_b) \tag{5}$$

Several research works [15-17] showed that MTC is one of the important parameters affecting FAC and the experimental data are often expressed in terms of Sherwood (Sh), Reynolds (Re) and Schmidt (Sc) numbers as:

$$Sh \ = \ a.Re^b.Sc^c \tag{6}$$

where a, b and c are related to mass transfer which occurs under a given flow condition and can only be obtained experimentally. Where (Sh) in the non-dimensional representation of MTC as a function of the local hydrodynamic parameters and expressed as:

$$Sh \ = \ MTC.d_H \ / \ D \tag{7}$$

where: d_H = hydraulic diameter, and D = diffusion coefficient of iron in water.

In Equation (6), the velocity exponent varies between 0.8 for lower Reynolds numbers and 1.0 for very high Reynolds numbers. This difference in the velocity exponent is caused by the surface roughness. This indicates that the FAC rate increases as the surface roughness increases. It should be also noted that the experimental studies and the correlations developed for MTC were carried out under low flow rates conditions compared with common operating conditions in power generation industry. Therefore, the MTC data obtained in the literature for moderate and high Reynolds numbers at power plant conditions can lead to significant errors.

In the case of piping downstream orifices, Tagg et al. [18] described the wear enhancement profile empirically in terms of the maximum Sherwood number using Reynolds number at the vena contracta section (Re_o) as follows:

$$Sh_{max} = 0.27 \cdot Re_o^{0.67} \cdot Sc^{0.33} \tag{8}$$

On the other hand, the local enhancement profile downstream of the orifice at different axial locations (z) is described empirically by Coney [19] referred to by Chexal et al. [20] as:

$$\frac{Sh_z}{Sh_{fd}} = 1 + A_z \left[1 + B_z \left(\frac{Re_o^{0.66}}{0.0165 \cdot Re^{0.86}} - 21 \right) \right] \tag{9}$$

where Sh_{fd} is the Sherwood number for the smooth straight pipe, Sh_z is the Sherwood number at any axial location (z), A_z and B_z are empirical constants.

Another modelling approach for FAC prediction was carried by Remy et al., [21], to develop a criteria in order to avoid FAC damage. They model the FAC mechanism as a combination of two main mechanisms. These mechanisms are divided into two steps. In the first of which is the production of soluble ferrous ions, which is represented by a first-order reaction:

$$V_C = \ K \ (C_{eq} - C) \tag{10}$$

V_C = total corrosion rate, K = reaction rate constant, C_{eq} = the soluble ion concentration at oxide water interface. The second step of FAC is correlated to the transfer of ferrous ion into the bulk water, which is a convective transport phenomenon that can be modelled as:

$$F_{IF} = \ k \ (C - C_\infty) \tag{11}$$

where: F_{IF} = the ferrous ion flux, MTC = mass transfer coefficient, C_∞ = ferrous ion concentration in the bulk flow. These two steps are assumed to be equal at equilibrium state, and can be expressed in terms of the combined equation:

$$V_C = \ 2K(MTC) \ (C_{eq} - C)/(MTC + 2K) \tag{12}$$

which was finally reduced to:

$$T_L = \ kC_{eq} \tag{13}$$

T_L is the thickness loss kinetic which is mainly correlated to the FAC wear rate. This conclusion of linear relationship between FAC rate and concentration was early refuted by Poulson [22], giving reasons in favour of non-linearity. These reasons including:

a.   removal of a surface film above a critical value of K, e.g. carbon steel in nitrate solutions,
b.   interactions of anodic and cathodic areas, e.g. iron in Nacl or $FeCl_3$ solutions,
c.   coupling of reactions, where flow effects K, which changes the corrosion potential and also the oxide solubility (ΔC). This leads to a dependency on K^n (where n is between 1 and 3), and
d.   dual control, e.g. situations in which the rate is partially controlled by activation such as copper alloys in seawater, or alternatively by two transport processes. This tends to lead to a dependency on K^n, where n is less than 1.

A comprehensive report on FAC in power plants, by EPRI [3], presents empirical models that have been used successfully to predict components that are most likely to wear, and provide reasonable estimates of pipe wall-thinning. These models are however computer-based due to the large amount of information that needs to be processed. These models are represented in the following two groups:

i.   Berge model

The model assumed that the chemical dissolution of the magnetite at the oxide/water interface occurs in accordance with the following equation:

$$DR = \ K \ (C_{eq} - C_S) \tag{14}$$

where DR is the ferrous iron production rate; K is the reaction rate constants or (kinetics); Ceq, is the equilibrium concentration of magnetite; and Cs, is the magnetite concentration at oxide/water interface. Finally, the FAC rate is given as:

$$FAC \ \text{rate} = 2K \ (C_{eq} - C_S) \tag{15}$$

ii.   MIT model

This model is mainly an improved version of Berge's model considering the diffusion through the porous oxide layer and incorporating both oxide thickness (δ) and porosity (Θ), and also considers diffusion of iron hydroxides through the pores, to determine the FAC rate as:

$$FAC \text{ rate } = \frac{\theta(C_{eq} - C_\infty)}{\left(1/K^* + (1-f)(1/k + \delta/D)\right)} \tag{16}$$

3. FAC modelling in two-phase flow

For FAC under two-phase flow condition, Remy et al. [21] indicated that same correlations used for single-phase flow calculations are also applicable in two-phase flow calculations. The only difference in the analysis is the calculation of actual Reynolds number, using the actual water velocity taking into account the void fraction between the steam and water.

EPRI report [3] incorporated the idea of Remy et al. [21] on the use of single phase correlations in clculating FAC under two-phase flow conditions. Calculation begin by determining Reynolds number for the liquid phase, Re_L, as follows:

$$Re_L = V_L \frac{d_H}{v_L} \tag{17}$$

Where the liquid velocity expressed as:

$$V_L = \left(\frac{Q}{A\rho_L}\right) \cdot \left(\frac{1-x}{1-\alpha}\right) \tag{18}$$

The flow is considered to be a single phase liquid flow when the steam quality (x) and the steam void fraction (α) are both equal to zero (0), while $\alpha = 1$ implies no presence of liquid, hence $Re_L = 0$ and no FAC damage is expected. Under two-phase flow conditions, when (α) is greater than (x), and $Re_L > Re$, the mass transfer coefficient increases and consequently FAC wear rate increases.

In two-phase flow the void fraction is usually not equal to steam quality because the liquid and the vapour phases are moving with different velocities, and sometimes in different directions. If homogeneous flow is assumed, the two phase's velocities are assumed to be equal, which results in a steam quality-void fraction relationship as a function of pressure. However, due to the fact that the homogeneous flow assumption is not suitable for many industrial applications, more sophisticated models were developed by researchers to relate steam quality to void fraction. One of the most cited model is the Chexal-Lellouche void model:

$$\langle \alpha \rangle = \frac{\langle j_g \rangle}{C_0 \langle j \rangle + \overline{\overline{V_{gj}}}} \tag{19}$$

where:

$\langle j \rangle$ and $\langle j_g \rangle$ are the mixture and vapour volumetric fluxes.

Another model developed by Kuo-Tong et al. [23] to predict FAC damage locations on High pressure (HP) turbine exhaust steam line. Their choice of High pressure (HP) turbine exhaust steam line as a case study was based on the plant measured data of pipe thickness which indicated HP lines as a good example where serious FAC takes place under two phase flow conditions. They reported that FAC phenomenon strongly depends on the piping layout and local flow conditions. They proposed a new mathematical approach to simulate FAC wear rate. The approach includes the use of 3D two-phase flow hydrodynamic CFD model to simulate the two-phase flow behaviour in HP lines, integrated with FAC Models to investigate the impact of the local parameters on FAC damage. The improvement of their new approach over the previous work is the ability to account for the multi-dimensional characteristics applied to FAC wear rate prediction. This is consider a great advantage in predicting FAC compared to the previous codes such as CHECKWORK or CAECE program which are based on empirical correlations that are dependent on the global flow conditions in the piping lines. In developing their code, the following assumptions were adopted:

- Adiabatic flow, since the piping in the type of the above-cited facilities is generally thermally insulated.
- Droplet-type two-phase flow, because of the high steam quality (> 85%) usually experienced in the studied piping. Hence, the vapour phase (steam) was modelled as a continuous phase and the liquid droplet as a dispersed phase.
- Simplified geometries, where valves in the piping are not considered.

Their final hydrodynamic CFD models include two-fluid 3D continuity and momentum equations. Closure relations such as the mixture k-e turbulent models, two-phase constitutive equations were also used. The developed FAC model claimed to include droplet impingement model, which was used to simulate the mechanically-assisted form of material degradation.

In a similar two-phase flow analysis applied to the extraction piping system connecting the low-pressure turbine (LPTB) and feed water heater (FWH) at boiling water reactors (BWR), Yuh et al. [24] developed a code to predict FAC wear in BWR piping system. Their mathematical approach is very similar to the work done by Kuo-Tong et al. [23], with the exception to the addition of corrosion model used. They obtained the local distributions of fluid parameters including two-phase flow velocities, void fractions, turbulent properties, and pressure. These results were further used to establish the relation for droplet kinetic energy that represents the FAC damage. The comparison of between their model and the plant measured data show qualitatively a good agreement.

4. FAC prediction downstream an orifice

Once the relationship between mass transfer and FAC wear rate is established, the computational model for MTC downstream of an orifice can be formulated. Fully developed turbulent pipe flow is assumed in order to determine MTC profiles downstream of the orifice. ANSI specifications of orifice were used to construct the geometrical model. Since the experimental condition in the present study is carried out for straight pipe section fabricated from hydrocal (CaSO4.½H2O) downstream of an orifice to simulate faster effect of mass transfer rates. The Solution is obtained for Renormalization Group (RNG) K-$\mathcal{E}$ differential viscosity model for turbulent flow in conjunction with the species transport equations using FLUENT CFD code.

The velocity field of the incompressible viscous flow is obtained using the one dimensional Reynolds averaged governing equations as follows:

Continuity equation:

$$\frac{\partial \bar{u}_i}{\partial x_i} = 0 \quad \left(\frac{\partial u'_i}{\partial x_i} = 0 \right) \tag{20}$$

Momentum equation:

$$\bar{u}_j \frac{\partial \bar{u}_i}{\partial x_j} = -\frac{\partial \bar{P}}{\partial x_i} + \frac{\partial}{\partial x_j}\left(\frac{1}{Re}\frac{\partial \bar{u}_i}{\partial x_j} - \overline{u'_i u'_j} \right) \tag{21}$$

Species mass transport equation for a steady process with no chemical reaction is:

$$\nabla.(\rho \vec{v} Y_i) = -\nabla.\vec{J}_i + S_i \tag{22}$$

where: $\vec{J}_i$ is the diffusion flux of species i, and arises due to concentration gradient, and S_i is the source term.

In Equation (11), the Boussinesq eddy viscosity assumption [25] is used for modeling the Reynold's stress. The eddy viscosity model relation is expressed as:

$$-\rho \overline{u_i u_j} = \mu_t [(\partial U_i / \partial z_j) + (\partial U_j \partial z_i)] - 2\rho k \delta_{ij} / 3 \tag{23}$$

where μ_t is defined as the "turbulent viscosity" and expressed as:

$$\mu_t = C_\mu f_\mu (\rho k^2 / \varepsilon) \tag{24}$$

The turbulence kinetic energy (k) and the turbulence kinetic energy dissipation rate (ε) are defined as follows:

$$k = (\overline{u^2} + \overline{v^2} + \overline{w^2}) / 2, \ \varepsilon = v\overline{(\partial u_i / \partial z_j)^2} \tag{25}$$

Therefore, the equation for turbulence kinetic energy can be also expressed as follows:

$$\frac{\partial(\rho U k)}{\partial z} + (1/r)\frac{\partial(r\rho V k)}{\partial r} = \frac{\partial[(\mu_{eff} / \sigma_k)(\partial k / \partial z)}{\partial z} + [1/r]\frac{\partial[(r\mu_{eff} / \sigma_k)(\partial k / \partial r)]}{\partial r} + G_k - \rho\varepsilon \tag{26}$$

and the equation for the turbulence kinetic energy dissipation expressed as:

$$\frac{\partial(\rho U \varepsilon)}{\partial z} + (1/r)\frac{\partial(r\rho V \varepsilon)}{\partial r} = \frac{\partial[(\mu_{eff}/\sigma_\varepsilon)(\partial\varepsilon/\partial z)]}{\partial z} + [1/r]\frac{\partial[(r\mu_{eff}/\sigma_\varepsilon)(\partial\varepsilon/\partial r)]}{\partial r} + \\ (\varepsilon/k)(C_{\varepsilon1}f_1 G_k - C_{\varepsilon2}f_2\rho\varepsilon)$$ (27)

In Equations (16) and (17) the generation of kinetic energy of turbulence term (G_k) can be written as:

$$G_k = \mu_{eff}[2[(\partial U/\partial z)^2 + (\partial V/\partial r)^2 + (V/r)^2] + [(\partial U/\partial r) + (\partial V/\partial z)]^2]$$ (28)

where the effective viscosity (μ_{eff}) is defined as:

$$\underbrace{\mu_{eff}}_{effective} = \underbrace{\mu}_{molecular} + \underbrace{\mu_t}_{turbulent}$$ (29)

The calculation of the local MTC is obtained similar to El-Gammal et al. [26] as:

$$MTC(z) = \frac{-D_{SL}\partial c/\partial n|_w}{(c_w - c_b)}$$ (30)

where c_w is the species concentration along the wall (obtained from hydrocal properties table), c_b is the species concentration in the bulk flow beyond the diffusive boundary layer, n is the normal vector to the wall surface and D_{SL} is the diffusive coefficient of the solid species, which is calculated by using Wilkie's semi-empirical relationship, [26]:

$$D_{SL} = \frac{7.4\times10^{-15}\times T\times\sqrt{\Psi\times M_S}}{\eta\times V^{0.6}}$$ (31)

where T is the temperature (K), Ψ is the association factor for the solvent (2.6 for water), M_S is the molecular mass for the solvent (18 g for water), η is the solvent absolute viscosity (Pa.s), V is the molecular volume of the dissolved species (144.86cm³/mol for hydrocal) at ambient temperature.

The concentration of hydrocal species in the bulk of water c_b is calculated as follows:

$$c_b(z) = \frac{1}{\rho U A}\int\rho u(r)c(r)dA$$ (32)

where U is the area average velocity, $u(r)$ is the instantaneous flow velocity, $c(r)$ is the species concentration profile, and A is the cross sectional area of the pipe. The term $\partial c/\partial n|_w$ (z) is calculated by taking the concentration gradient at the wall at an axial location (z). Substituting equations (31) and (32) into equation (30), the cross-sectional average for $MTC(z)$ along the axial direction can be calculated.

In order to evaluate the FAC wear rate downstream orifice, the conservation equations are integrated over each control volume in the flow field which extended upstream downstream of the orifice. Reynolds Average Navier-Stokes equations were solved using the K-ε (RNG) differential viscosity turbulence model to account for low-Reynolds-number (LRN) effects. The solution convergence is greatly improved by using fine near-wall grids. Also, the computational mesh was refined where high velocity and species concentration gradients were expected. Due to the high Schmidt numbers encountered in mass transfer problems, Nesic et al. [27] suggested that the maximum value of Y^+ should not exceed 0.1.The Reynolds average mass transport equation was also solved for determining the concentration field of the dissolved wall species.

For the present case study, numerical simulations were performed at Reynolds number, Re = 20,000 and orifice-to-pipe diameter ratios of d/D = 0.25, 0.5 and 0.74. Prior to the commencement of the simulations, a sensitivity study of three different grid numbers was performed and the results indicate a deviation in flow characteristics within ±2%. The flow characteristics for the three orifice geometries are found to be qualitatively similar. Therefore, only representative vector and contour plots for d/D = 0.5 are presented here. Fig. (2a) shows the mean velocity vectors normalized by the averaged inlet velocity (U_o) within

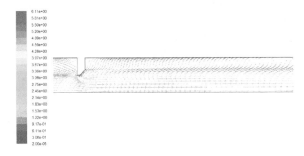

a) Normalized mean velocity vectors downstream the orifice

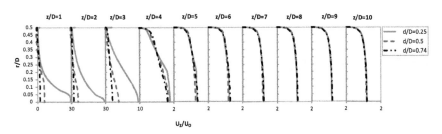

b) Profiles of normalized horizontal velocity component (Re = 20,000 d/D = 0.25, 0.5 and 0.74)

Figure 2. Numerical results for the flow downstream orifice

the flow domain. It can be seen that the flow accelerates as it approaches the orifice then separates at the sharp edges of the orifice, forming large vortices downstream. These vortices sustain the reduction in the flow cross-sectional area further downstream up to the minimum area known as the vena contracta, after which the flow decelerates towards the flow reattachment point. This is the cause of the high velocity central region observed just downstream the orifice which changes to lower velocity region as the flow develops further downstream. Fig. (2b) shows the profiles of the mean horizontal velocity for the three geometries at different axial locations. The ordinate r/D is measured from the centreline of the pipe while Z/D is measured from the orifice. As shown in the figure, the maximum centreline velocity increases within the circulation zone as the orifice diameter decreases. The relative reductions in the centreline velocity at Z/D=1 through Z/D=4 are about 93%, 92%, 75% and 5.7% respectively, as d/D increases from 0.25 to 0.74. Downstream Z/D = 5, the velocity profiles are almost similar for the three geometries, with the flow returning to fully developed turbulent flow at $Z/D \cong 30$. The flow reattachment length downstream the orifice also increases as the orifice diameter decreases, with the shortest at $Z/D \cong 2$ for d/D = 0.74, or by $Z/D \cong 3$ for d/D = 0.5 and $Z/D \cong 4$ for d/D = 0.25.

Another important flow parameter that can be related to the FAC is the skin friction coefficient C_f which is defined as $C_f = \tau w/(1/2\rho U_0^2)$, where τw is the wall shear stress. The distribution of C_f downstream the orifice for the three d/D ratios are illustrated in Fig. (3). Variation of C_f is found to be similar for the three d/D ratios. Typical profile show an increase steeply downstream the orifice due to the reversed flow generated by the separating vortices, and reaches a maximum value at $Z/D \cong 0.4$, 1.3 and 2.3, for d/D = 0.74,

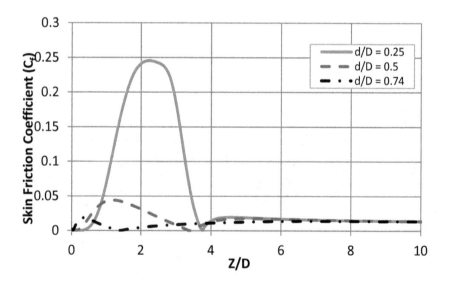

Figure 3. Skin friction coefficient along the pipe wall downstream the orifice for different orifice diameters

0.5 and 0.25 respectively. The skin friction coefficient then decreases steeply and reaches a minimum value at $Z/D \cong 1.4$, 3.4 and 3.7, for $d/D = 0.74$, 0.5 and 0.25 respectively, within the flow re-attachment region. As the flow progresses downstream, the surface shear stress increases due to the boundary layer developed by the reattached flow, and reaches the second local peak between $Z/D = 4$ and 5. These local peak values of C_f decrease as d/D increases, the first value decreases by about 92% as d/D increases from 0.25 to 0.74, while the second value decreases marginally. In general, the shear stress found to decrease as the boundary layer thickness increases downstream restricting orifices.

The contours of normalized turbulent kinetic energy (TKE) within the flow domain for $d/D = 0.5$ are shown in Fig. (4a). TKE increases appreciably within the flow separation zone due to the high velocity gradients within this region. Fig. (4b) shows the profiles of the normalized

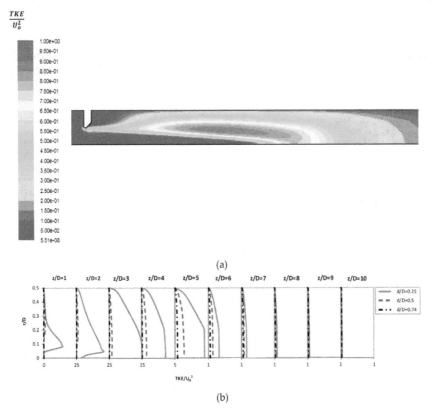

Figure 4. (a) – Contours of normalized turbulent kinetic energy downstream the orifice, for d/D = 0.5 and Re = 20,000 and (b) radial profiles of normalized turbulent kinetic energy at Re = 20,000.

TKE for the three d/D at different axial locations downstream the orifice. The profiles are found to be qualitatively similar for all the three d/D; within the circulation region ($Z/D = 0 –$

2). The normalized *TKE* value increases from zero at the wall to a maximum value at the centreline. These local peak values of *TKE*, as well as its radial location from the wall, increase as d/D decreases. All the profiles however collapse to a single line at $Z/D = 8$ as the flow becomes fully developed. The axial distribution of normalized *TKE* downstream the orifice is shown in Fig. (4b) for the three d/D ratios. Moreover, the peak value of *TKE* decreases by about 96% as d/D increases from 0.25 to 0.74.

The predicted *MTC* distributions downstream the orifice is shown in Fig. (5). The mass transfer found to increase sharply downstream the orifice and reaches a peak value at $Z/D \cong$ 1, 4 and 3, for $d/D = 0.74$, 0.5 and 0.25 respectively, after which it decreases steeply to the fully developed value downstream. The *MTC* distributions found to correlate very well with the *TKE* profiles. The non-dimensional mass transfer coefficient represented by Sherwood number (*Sh*) and the mass transfer enhancement *ShzShfd* downstream the orifice are shown in Figs. (6a and 6b). The peak value of *Sh*, as well as *ShzShfd*, decreases by about 63% and 42% when d/D increases from 0.25 to 0.5 and from 0.5 to 0.74 respectively. The axial locations of the peak values however move downstream from $Z/D \cong 1$ to 4 when d/D decreases from 0.74 to 0.5, while the peak locations move upstream from $Z/D \cong 4$ to 3 when d/D decreases from 0.5 to 0.25.

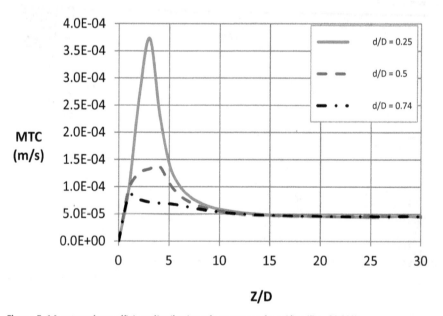

Figure 5. Mass transfer coefficient distributions downstream the orifice ($Re = 20,000$)

FAC rate downstream the orifice is shown in Fig. (7) for different d/D ratios. The peak value of *FAC* increases as d/D decreases. In the present analysis, the peak value increases by about 90% when d/D decreases from 0.74 to 0.25. Moreover, the axial locations of the peak values however move downstream from $Z/D \cong 1$ to 3 as d/D decreases from 0.74 to 0.25. The location of FAC peak values found to also correlate very well with the location of the peak TKE.

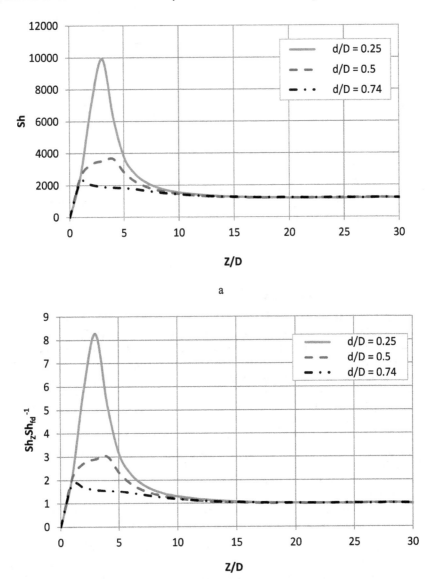

Figure 6. a: Sherwood number distributions downstream the orifice, for different orifice diameters, and $Re = 20,000$, b: Enhancement of mass transfer downstream the orifice, for different orifice diameters, and $Re = 20,000$

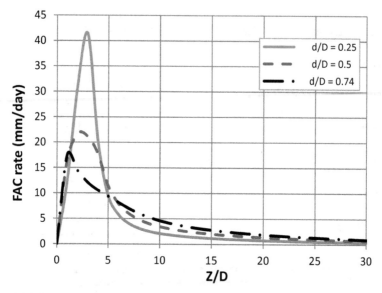

Figure 7. FAC wear rate downstream the orifice, for different orifice diameters, and $Re = 20,000$

4.1. Power plants inspection data

Ultrasonic techniques (UT) measurements are commonly used to determine the wall thinning measurement in nearly all power plants and to provide more accurate data for measuring the remaining wall thickness in piping system. The UT inspection data were obtained at grid intersection points marked on the piping component. The data are usually stored in a data logger and transferred to a PC for further processing using appropriate software. The wall thickness data were obtained at different grid point on the piping as shown in Fig. (8). For each pipe downstream an orifice, the difference between the measured wall thickness and the nominal pipe wall thickness is calculated and considered to be the wear at this axial location along the pipe. Sometimes, scanning within grids and recording the minimum found within each grid square is an acceptable alternative to the above method. However, it should also be noted that scanning within grids and the minimum wall thickness recorded can affect the accuracy of the data if point-to-point comparison between two consecutive inspections times. The inspection data are used to determine whether the component has experienced wear and to identify the location of maximum wall thinning as well as to evaluate the wear rate and indentify wear pattern in piping component.

In the present case study, 132 inspection data collected from 5 nuclear power plants and 3 fossil power plants for piping downstream an orifice were analyzed. The data of very high and low values of wear are compared to adjacent inspection readings in order to remove data outliers. Once the data set for each inspection location is verified, the wear is identified at each band along the pipe axis. The measured wear data at different location from the orifice were presented for different piping systems as shown in Fig. (9). It can be concluded that the effect of geometry found to strongly affect the FAC wear rate downstream of the orifice. The maximum value of the wear found to be located within 5D downstream of the orifice which agrees very well with the inspection data collected from different power plants. Also, the hydrodynamic profiles such as *TKE* and *MTC* found to characterize the FAC wear rate downstream the orifice and the value of the maximum FAC wear rate increases as the orifices diameter reduces.

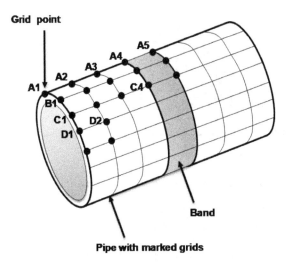

Figure 8. Inspection grids downstream of the orifice

5. Effect of proximity between piping components on FAC

From the above case study of the flow downstream an orifice, it can be concluded that the size and the geometry of a piping component directly influence the flow velocity and hence the local mass transfer rate. In addition, components with geometries that promote increase in the velocity and turbulence tend to experience higher FAC rate such as elbow, tee, reducers and valves, etc. The effect of turbulence on the FAC rate is represented by the geometry enhancement factor as described by Chexal et al. [20]. Moreover, a component that has another component located close upstream experience more turbulence that further increases the FAC rate. One would expect that such components tend to experience more severe FAC as discussed by Kastner et al. [28] and Poulson [29]. In fact, the piping geometry at the point of rupture at Surry 2, where the elbow located downstream of a T-fittings, is a strong evidence of the proximity effect.

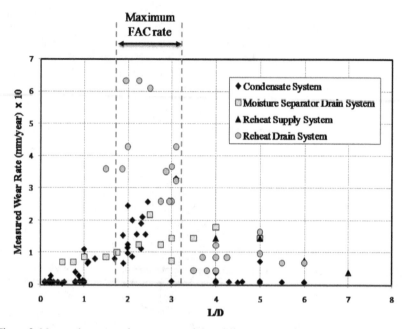

Figure 9. Measured wear rate downstream orifces at different power plants

In summary, a significant amount of research has been conducted on investigating the effect of fluid chemical properties on FAC in nuclear power plants, online monitoring for strategic FAC locations, and FAC control based on chemical fluid decomposition. However, the hydrodynamic effects of single- and two-phase flows on FAC have not been thoroughly investigated, and are currently not well understood. The present case study aims to evaluate the effect of components located in proximity to another component upstream on the pipe wall-thinning rate due to FAC using real inspection data from different degraded elbows in several power plants. Also, to inspire future discussions among the scientific community to revisit the effect of component geometry on FAC wear rate in order establish new codes/correlations that meet actual industrial findings. In order to accomplish this goal, ultrasonic (UT) inspection data for 90° carbon steel elbows for different systems across different nuclear power stations are analyzed in order to quantify the effect of the proximity between components on the FAC wear rate. The effect of the flow velocity as well as the distance between the elbows and the upstream components is also discussed in the present analysis.

The increase in the wear rate due to the proximity is calculated for 90° elbows located within 1D of the upstream Components. Ahmed [4] evaluated several UT data of back-to-back elbows for deferent nuclear power stations as shown in Figure (10). The results are clearly showing an increase in the wear rate due the proximity of the elbow with the upstream components as the majority of the data are consistently above zero. The increase in the wear rate is approximately equal to 70% in average. The distribution of the data occurrence is plotted in Figure (11) to represent the percentage of components to the total used inspections that have the same wear rate. If a linear wear over the operating time is assumed, one can determine the remaining life of component located in the proximity from another component will results in reduction in component life by more than half life. Although the results show a large scatter of the data with a standard deviation of 50%, the proximity is believed to have a significant effect on the wear rate.

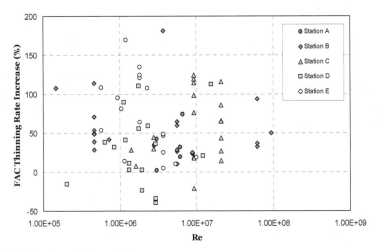

Figure 10. Effect of proximity on FAC wear rate

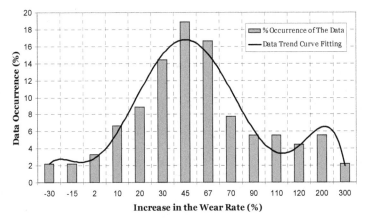

Figure 11. Data occurrence

The other set of UT data is collected for 90° elbows located at different distances (L) from the upstream component are presented. The ratio (L/D) is used to represent the non-dimensional upstream distance with respect to the pipe diameter. As discussed before, a component that has another component located close upstream is expected to experience more turbulence which increases the FAC rate. On the other hand, it is expected that the effect of the turbulence produced from the upstream component becomes much less as the distance from the upstream component increases as shown in Fig. (12). It should be noted that the extent of the proximity effect as shown in Figure (6) is 0 to 5 piping diameters as the a average increase in the wear rate is equal to 70 % (with a standard deviation of 50%) which approximately the same as for elbows within 1D from the upstream components. As the distance between the two components increases the change in the wear rate decreases to reach a minimum value in the fully developed region.

A general trend between the increase in the wear rate and the non-dimensional upstream distance (L/D) is obtained using the average values in the wear rates at different (L/D). The additional effect of the component located upstream close to another component is previously correlated by Kastner et al. (1990) as:

$$\text{Effect of Proximity } (\%) \ = e^{\left(-0.231 \cdot L/D\right)} \times 100 \tag{33}$$

The average values of FAC wear rate in Fig. (12) are compared to the empirical equation proposed by Kastner et al. [28] and found to be in good agreement with the present data shown in Fig. (13). However, the correlation tends to over predict the data as the distance between the components approaches zero which can be attributed to the accuracy of the proposed correlation and the UT measurements uncertainty.

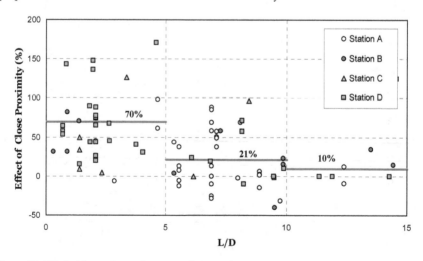

Figure 12. Effect of the upstream distance on the wear rate

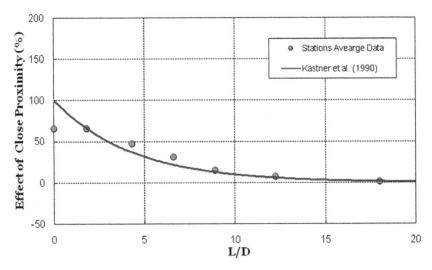

Figure 13. Comparing data with Kastner et al. [28] correlation

As discussed before, components with geometries that promote increase in the flow velocity and turbulence tend to experience higher FAC rate such as elbow, tee, reducers downstream of orifices and valves, etc. Therefore, the turbulence production is expected to be dependent on the component geometry. Consequently, the type of the upstream component could be also considered as a contributing factor and should be considered while performing the flow and mass transfer analysis.

6. Conclusions and recommendations

The effect of geometry found to strongly affect the FAC wear rate. For example, the maximum value of the wear found to be located within 5D downstream of the orifice. Also, the analyzed trends obtained from the plants inspection data show significant increase in the wear rate due to the component proximity. The CFD simulation of the hydrodynamic parameters such as *TKE* and *MTC* found to characterize the FAC wear rate downstream of piping components. Additional experimental and numerical investigations are required to further evaluate the close proximity effect for different components configuration. Findings of the present work will allow the power plant FAC engineers to identify more accurately susceptible components and reduce the inspection scope/time.

Author details

Wael H. Ahmed

Department of Mechanical Engineering, King Fahd University of Petroleum & Minerals, Dhahran, Saudi Arabia

Acknowledgement

The support provided by the Deanship of Scientific Research (DSR) at King Fahd University of Petroleum & Minerals (KFUPM) for funding this research work through project No. IN090038, is gratefully acknowledged. The author also thanks Mr. Mufatiu Bello for carrying out the data analysis of the orifice case study.

Also, the author would like to thank Atomic Energy of Canada Ltd. (AECL), and the CANDU Owner Group (COG) for their support and permission to publish the data on plant close-proximity effect presented in this chapter. The data have been also presented by the author and analyzed in the Annals of Nuclear Energy (37 (2010) 598–605)

7. References

[1] Kanster, W., Erve, M., Henzel, N., and Stellwag, B., (1990), "Calculation code for erosion corrosion induced wall thinning in piping system, Nuclear Engineering and Design, Vol. 119, pp. 431-438.

[2] Yurmanov, V., Rakhmanov, A., (2009), "Workshop on erosion-corrosion, International Atomic Energy Agency, Workshop on Erosion-Corrosion, Moscow, Russian Federation.

[3] Moore, F.E., (2008), "Welding and repair technology for power plants, 18th Int. EPRI Conference.

[4] Ahmed, W.H. (2010), "Evaluation of the proximity effect on flow accelerated corrosion", Annals of Nuclear Energy, Vol. 37, pp. 598-605.

[5] Crawford, N.M., Cunningham, G., Spence, S.W.T. (2007), "An experimental investigation into the pressure drop for turbulent flow in 90° elbow bends" Proc. of the Institution of Mechanical Engineers, Part E, Journal of Process Mechanical Engineering, Vol. 221, no. 2, pp. 77-88.

[6] Poulson B. (1999), "Complexities in predicting erosion corrosion", Wear 233-235, pp. 497-504

[7] Chen X, McLaury B.S., Shirazi S. A.,, (2006), "A Comprehensive Procedure to Estimate Erosion in Elbows for Gas/Liquid/Sand Multiphase Flow " ASME Journal of Energy Resources Technology, Vol. 128, pp. 70-78.

[8] Kim S., Park J.H., Kojasoy G., Kelly J.M., Marshall S.O. (2007), "Geometric effects of 90-degree Elbow in the development of interfacial structures in horizontal bubbly flow ", Nuclear Engineering and Design, Vol.237, 2105-2113

[9] Hassan YA, Schmidl W., Ortiz-Villafuerte J. (1998), "Investigation of three-dimensional two-phase flow structure in a bubbly pipe flow" Measurement Science Technology, Vol. 9, 309-326

[10] Jepson, W. P., (1989), " Modelling the transition to slug flow in horizontal conduit", Canadian Journal of Chemical Engineering, Vol. 67, pp. 731-740

[11] Keating, A., and Nesic, S., (1999), " Prediction of two-phase erosion-corrosion in bends", Second International Conference on CFD in the Mineral and Process Industries, CSIRO, Melbourne, Australia, Dec. 6-8.

[12] Bozzini B., Ricotti M.E., Boniardi M., Mele C. (2003), "Evaluation of erosion-corrosion in multiphase flow via CFD and experimental analysis" Wear, Vol. 255, no. 1-6, pp. 237-245

[13] Chang Y.S., Song K.H., Lee S.M., Choi J.B., Kim Y.J. (2006), Key Engineering Materials, Vol. 321-323, pp. 670-3

[14] Ferng, Y.M., (2008), " Prediction local distribution of erosion-corrosion wear sites for the piping in the nuclear power plant using CFD models" Annals of Nuclear Energy, Vol. 35, no. 2, pp. 304-313.

[15] Poulson B. (1987), "Predicting the occurrence of Erosion-corrosion, Plant corrosion: prediction of material performance, Strutt, J.E., and Nicholls, J. R., Editors, Ellis Horwood Ltd.

[16] Berge, P., and Saint Paul, P., (1981), "Water Chemistry of Nuclear Reactor Systems" Proceedings of the British Nuclear Energy Society, London, P. 19.

[17] Bouchacourt, M., and Remy, F. N., (1991), Proceeding of 3rd NACE international Region Management Committee Symposium, Cambridge, U.K.

[18] Tagg, D. J., Patrick, M. A., and Wragg, A. A., (1979), "Heat and Mass Transfer Downstream of Abrupt Nozzle Expansions in Turbulent Flow "Trans. Inst. Chem. Eng., Vol 57, no 12, pp.176-181.

[19] Coney, M., (1980), CERL Internal Report, Ref: RD/L/N197/80.

[20] Chexal, B., Horowitz, J., Jones, R., Dooley, B., Wood, C., Bouchacourt, M., Remy, F., Nordmann, F., and St. Paul, P., 1996 "Flow-Accelerated Corrosion in Power Plants," Electric Power Research Institute Report No. TR-106611.

[21] Remy, F.N., Bouchacourt, M., 1992, "Flow-assisted corrosion: a method to avoid damage. Nuclear Engineering and Design", 133: p. 23-30.

[22] Poulson, B., 1993,"Advances in understanding hydrodynamic effects on corrosion", Corrosion Science, 35: p. 655-665

[23] Kuo-Tong, M.A., et al. 1998, "Numerically Investigating The Influence Of Local Flow Behaviours On Flow-Accelerated Corrosion Using Two-Fluid Equations", Nuclear Technology, 123: p. 90-102.

[24] Yuh Ming Ferng et al., 2008, "A Two-phase Methodology To Predict FAC Wear Sites In The Piping System of A BWR", Nuclear Engineering and Design, 238: p. 2189-2196.

[25] Boussinesq, J. (1877), "Théorie de l'Écoulement Tourbillant", Mem. Présentés par Divers Savants Acad. Sci. Inst. Fr., Vol. 23, pp. 46-50.

[26] El-Gammal M., Mazhar H., Cotton J.S., Shefski C., Pietralik J., Ching C. Y., (2010), "The hydrodynamic effects of single-phase flow on flow accelerated corrosion in a 90-degree elbow", Nuclear Engineering and Design, vol. 240, 6, pp. 1589-159.

[27] Nesic, S., Adamopoulos, G., Postlethwaite, J. and Bergstrom, D.J.: "Modelling of Turbulent Flow and Mass Transfer with Wall Function and Low-Reynolds Number Closures". The Canadian Journal of Chemical Engineering, Vol. 71, pp. 28-34.

[28] Kastner W., Erve M., Henzel N., and Stellwag B., (1990), Nuc. Eng. Design, v. 119, pp. 431-438.

[29] Poulson, B. (2007), Proc. of 13th Int. Conf. On Environmental Degradation of Materials in Nuclear Power Systems- Canadian Nuclear Soc, Whistler, B.C., Canada.

Chapter 3

Reliability of Passive Systems in Nuclear Power Plants

Luciano Burgazzi

Additional information is available at the end of the chapter

1. Introduction

In order to tackle the development of advanced nuclear technologies, the reliability of passive systems has become an important subject and area under discussion, for their extensive use in new and advanced nuclear power plants, (NEA, 2002), in combination with active safety or operational systems.

Following the IAEA definitions, [1], a passive component does not need any external input or energy to operate and it relies only upon natural physical laws (e.g. gravity, natural convection, conduction, etc.) and/or on inherent characteristics (properties of materials, internally stored energy, etc.) and/or 'intelligent' use of the energy that is inherently available in the system (e.g. decay heat, chemical reactions etc.).

The term "passive" identifies a system which is composed entirely of passive components and structures or a system which uses active components in a very limited way to initiate subsequent passive operation. That is why passive systems are expected to combine among others, the advantages of simplicity, a decrease in the need for human interaction and a reduction or avoidance of external electrical power or signals. These attractions may lead to increased safety and acceptability of nuclear power generation if the detractions can be reduced.

Besides the open feedback on economic competitiveness, special aspects like lack of data on some phenomena, missing operating experience over the wide range of conditions, and driving forces which are smaller - in most cases - than in active safety systems, must be taken into account: the less effective performance as compared to active safety systems has a strong impact on the reliability assessment of passive safety systems.

A categorisation has been developed by the IAEA in [1] distinguishing:

a.   physical barriers and static structures (e.g. pipe wall, concrete building).
This category is characterized by:
-   no signal inputs of "intelligence", no external power sources or forces,
-   no moving mechanical parts,
-   no moving working fluid.

Examples of safety features included in this category are physical barriers against the release of fission products, such as nuclear fuel cladding and pressure boundary systems; hardened building structures for the protection of a plant against seismic and or other external events; core cooling systems relying only on heat radiation and/or conduction from nuclear fuel to outer structural parts, with the reactor in hot shutdown; and static components of safety related passive systems (e.g., tubes, pressurizers, accumulators, surge tanks), as well as structural parts (e.g., supports, shields).

b.   moving working fluids (e.g. cooling by free convection).
This category is characterized by:
-   no signal inputs of "intelligence", no external power sources or forces,
-   no moving mechanical parts, but
-   moving working fluids.

Examples of safety features included in this category are reactor shutdown/emergency cooling systems based on injection of borated water produced by the disturbance of a hydrostatic equilibrium between the pressure boundary and an external water pool; reactor emergency cooling systems based on air or water natural circulation in heat exchangers immersed in water pools (inside containment) to which the decay heat is directly transferred; containment cooling systems based on natural circulation of air flowing around the containment walls, with intake and exhaust through a stack or in tubes covering the inner walls of silos of underground reactors; and fluidic gates between process systems, such as "surge lines" of Pressurized Water Reactors (PWRs).

c.   moving mechanical parts (e.g. check valves).
This category is characterized by:
-   no signal inputs of "intelligence", no external power sources or forces; but
-   moving mechanical parts, whether or not moving working fluids are also present.

Examples of safety features included in this category are emergency injection systems consisting of accumulators or storage tanks and discharge lines equipped with check valves; overpressure protection and/or emergency cooling devices of pressure boundary systems based on fluid release through relief valves; filtered venting systems of containments activated by rupture disks; and mechanical actuators, such as check valves and spring-loaded relief valves, as well as some trip mechanisms (e.g., temperature, pressure and level actuators).

d.   external signals and stored energy (passive execution/active actuation, e.g. scram systems).

This category addresses the intermediary zone between active and passive where the execution of the safety function is made through passive methods as described in the previous categories except that internal intelligence is not available to initiate the process. In these cases an external signal is permitted to trigger the passive process. To recognize this departure, this category is referred to as "passive execution/active initiation".

Examples of safety features included in this category are emergency core cooling and injections systems based on gravity that initiate by battery-powered electric or electro-pneumatic valves; emergency reactor shutdown systems based on gravity or static pressure driven control rods.

According to this classification, safety systems are classified into the higher categories of passivity when all their components needed for safety are passive. Systems relying on no external power supply but using a dedicated, internal power source (e.g., a battery) to supply an active component are not subject to normal, externally caused failures and are included in the lowest category of passivity. This kind of system has active and passive characteristics at different times, for example, the active opening of a valve initiates subsequent passive operation by natural convection.

Inclusion of failure modes and reliability estimates of passive components for all systems is recommended in probabilistic safety assessment (PSA)1 studies. Consequently the reliability assessment of passive safety systems, defined as the probability to perform the requested mission to achieve the generic safety function, becomes an essential step.

Notwithstanding that passive systems are credited a higher reliability with respect to active ones, – because of the smaller unavailability due to hardware failure and human error -, there is always a nonzero likelihood of the occurrence of physical phenomena leading to pertinent failure modes, once the system comes into operation. In fact the deviations of the natural forces or physical principles, upon which they rely, from the expected conditions can impair the performance of the system itself. This remark is especially applicable to type B passive systems (i.e. implementing moving working fluids) named thermal-hydraulic passive systems, due to the small engaged driving forces and the thermal-hydraulic phenomena affecting the system performance.

Indeed, while in the case of passive A systems the development of the structural reliability analysis methodology can be carried out with the application of the principles of the probabilistic structural mechanics theory, and operating experience data can be inferred for the reliability assessment of passive C and D components, there is yet no agreed approach as far as passive B systems are concerned.

In fact, such passive safety systems in their designs rely on natural forces, such as gravity or natural convection, to perform their accident prevention and mitigation functions once actuated and started: these driving forces are not generated by external power sources (e.g., pumped systems), as is the case in operating reactor designs. Because the magnitude of the natural forces, which drive the operation of passive systems, is relatively small, counter-forces (e.g. friction) can be of comparable magnitude and cannot be ignored as it is generally

---

[1] In the following PSA (Probabilistic Safety Assessment) and PRA (Probabilistic Risk Assessment) are utilized indifferently

the case of systems including pumps. Moreover, there are considerable uncertainties associated with factors on which the magnitude of these forces and counter forces depends (e.g. values of heat transfer coefficients and pressure losses). In addition, the magnitude of such natural driving forces depends on specific plant conditions and configurations which could exist at the time a system is called upon to perform its safety function. All these aspects affect the thermal-hydraulic (T-H) performance of the passive system.

Consequently, a lot of efforts have been devoted mostly to the development of consistent approaches and methodologies aimed at the reliability assessment of the T-H passive systems, with reference to the evaluation of the implemented physical principles (gravity, conduction, etc.). For example, the system fault tree in case of passive systems would consist of basic events, representing failure of the physical phenomena and failure of activating devices: the use of thermal-hydraulic analysis related information for modeling the passive systems should be considered in the assessment process.

The efforts conducted so far to deal with the passive safety systems reliability, have raised an amount of open issues to be addressed in a consistent way, in order to endorse the proposed approaches and to add credit to the underlying models and the eventual reliability figures, resulting from their application. In fact the applications of the proposed methodologies are to a large extent dependent upon the assumptions underlying the methods themselves. At the international level, for instance, IAEA recently coordinated a research project, denoted as *"Natural Circulation Phenomena, Modelling and Reliability of Passive Systems"* (2004-2008), [2,3], while another coordinated research project on *"Development of Methodologies for the Assessment of Passive Safety System Performance in Advanced Reactors"* (2008-2011) is currently underway: while focus of the former project has been the natural circulation and related phenomena, the objective of the latter program is to determine a common analysis-and-test method for reliability assessment of passive safety system performance. This chapter provides the insights resulting from the analysis on the technical issues associated with assessing the reliability of passive systems in the context of nuclear safety and probabilistic safety analysis, and a viable path towards the implementation of the research efforts in the related areas is delineated as well. Focus on these issues is very important since it is the major goal of the international research activities (e.g. IAEA) to strive to reach a common consensus about the different proposed approaches. The chapter is organized as follows: after an overview on passive safety systems being implemented in the design of innovative reactors and an introduction on the main components of Probabilistic Safety Assessment approach, at first the current available methodologies are illustrated and compared, the open issues coming out from their analysis are identified and for which one of them the state of the art and the outlook is presented; the relative importance of each of them within the evaluation process is presented as well.

2. Passive systems implementation in advanced reactor designs

Several advanced water cooled reactor designs incorporate passive safety systems based on natural circulation, as described in [2,3]: some of the most relevant design concepts for natural circulation systems are described hereafter and namely as regards AP600/AP1000, ESBWR and ABWR designs.

It is important to note that the incorporation of systems based on natural circulation to achieve plant safety and economic goals is being extended also to Generation-IV reactor concepts: however due to the early stage of the design - many systems are not yet established - they are not explicitly addressed.

2.1. AP600/AP1000 Passive Residual Heat Removal systems (PRHR)

Figure 1 presents a schematic that describes the connections of the primary system passive safety systems.

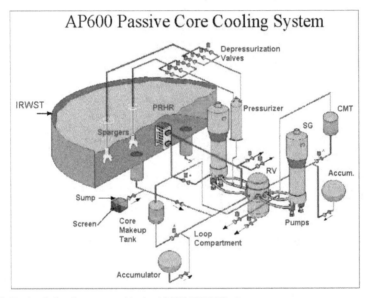

Figure 1. Passive Safety Systems used in the AP600/AP1000 Designs

The AP600/AP1000 passive safety systems consist of:

- A Passive Residual Heat Removal (PRHR) System
- Two Core Make-up Tanks (CMTs)
- A Four Stage Automatic Depressurization System (ADS)
- Two Accumulator Tanks (ACC)
- An In-containment Refueling Water Storage Tank, (IRWST)
- A Lower Containment Sump (CS)
- Passive Containment Cooling System (PCS)

The PRHR implemented in the Westinghouse AP1000 design consists of a C-Tube type heat exchanger in the water-filled In-containment Refuelling Water Storage Tank (IRWST) as shown in the schematic given in Figure 2. The PRHR provides primary coolant heat removal via a natural circulation loop. Hot water rises through the PRHR inlet line attached to one of the hot legs. The hot water enters the tube sheet in the top header of the PRHR heat exchanger at full system pressure and temperature. The IRWST is filled with cold borated water and is open to containment heat removal from the PRHR heat exchanger occurs by boiling on the outside surface of the tubes. The cold primary coolant returns to the primary loop via the PRHR outline line that is connected to the steam generator lower head.

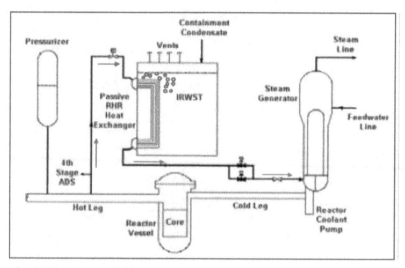

Figure 2. AP1000 passive residual heat removal systems (PRHR)

2.2. ESBWR (Economic Simplified Boiling Water Reactor) Isolation Condenser System (ICS)

During a Loss of Coolant Accident (LOCA), the reactor shuts down and the Reactor Pressure Vessel (RPV) is isolated by closing the main steam line isolation valves. The ICS removes decay heat after any reactor isolation. In other words, the ICS passively removes sensible and core decay heat from the reactor when the normal heat removal system is unavailable. Decay heat removal limits further increases in steam pressure and keeps the RPV pressure below the safety set point. The arrangement of the IC heat exchanger is shown in Figure 3.

The ICS consists of four independent loops, each containing two heat exchanger modules that condense steam inside the tube and transfers heat by heating/evaporating water in the IC pool, which is vented to the atmosphere. This transferring mechanism from IC tubes to the surrounding IC pool water is accomplished by natural convection, and no forced circulation equipment is required.

The ICS is initiated automatically by any of the following signals: high reactor pressure, main steam line isolation valve (MSIV) closure, or an RPV water level signal. To operate the

ICS, the IC condensate return valve is opened whereupon the standing condensate drains into the reactor and the steam water interface in the IC tube bundle moves downward below the lower headers.

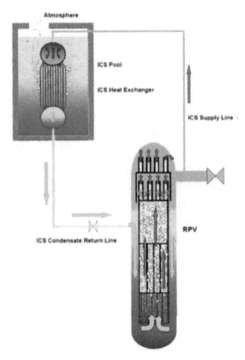

Figure 3. Isolation condenser arrangement

2.3. ESBWR Passive Containment Cooling System (PCCS)

The PCCS is a passive system which removes the decay heat released to the containment and maintains the containment within its pressure limits for design basis accidents such as a LOCA. The schematic of the PCCS is shown in Figure 4. The PCC heat exchangers receive a steam-gas mixture from the Dry Well (DW), condense the steam and return the condensate to the RPV via the Gravity Driven Cooling System GDCS pools. The non condensable gas is vented to the Wet Well (WW) gas space through a vent line submerged in the Suppression Pool (SP). The venting of the non condensable gas is driven by the differential pressure between the DW and WW. The PCCS condenser, which is open to the containment, receives a steam-gas mixture supply directly from the DW. Therefore, the PCCS operation requires no sensing, control, logic or power actuated devices for operation. The PCCS consists of six PCCS condensers. Each PCCS condenser is made of two identical modules and each entire PCCS condenser two-module assembly is designed for 11 MWt capacity. The condenser condenses steam on the tube side and transfers heat to the water in the IC/PCC pool. The evaporated steam in the IC/PCC pool is vented to the atmosphere. PCCS condensers are located in the large open IC/PCC pool, which are designed to allow full use of the collective water inventory.

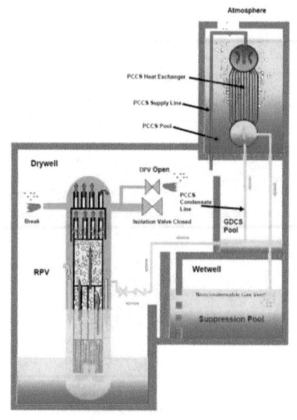

Figure 4. Passive containment cooling condenser arrangement

2.4. ABWR (Advanced Boiling Water Reactor) passive reactor cooling system and passive containment cooling system

The passive heat removal system (PHRS) consists of two dedicated systems (Figure 5, right) namely the passive reactor cooling system (PRCS: the same as Isolation condenser) and the passive containment cooling system (PCCS), that use a common heat sink pool above the containment allowing a one-day grace period, with a 4*50% redundancy (Figure 5, left). These passive systems not only cover beyond DBA condition, but also provide in-depth heat removal backup for the RHR.

In addition, they provide the overpressure protection safety function, practically excluding the necessity of containment venting before and after core damage. Figure 6 shows PCCS functional schematic and an example of containment pressure transient following typical low pressure core melt scenario.

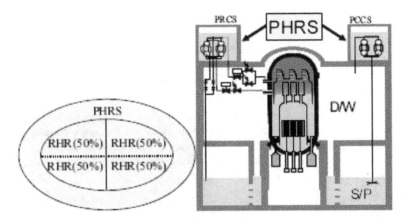

Figure 5. ABWR Passive heat removal system

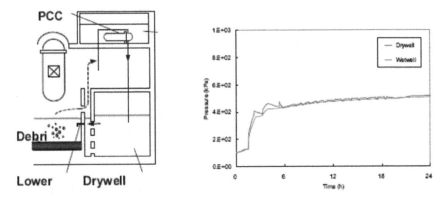

Figure 6. Example of containment pressure transient following typical low pressure core melt scenario.

3. Overview of PSA

PSA methodology widely used in the nuclear power industry is deemed helpful to the safety assessment of the facility and along the correspondent licensing process: probabilistic safety assessment can provide insights into safety and identify measures for informing designers of the safety of the plant.

The first comprehensive application of the PSA dates back to 1975, to the United States Nuclear Regulatory Commission's (U.S. NRC) Reactor Safety Study [4]. Since that pioneering study, there has been substantial methodological development, and PSA techniques have become a standard tool in the safety evaluation of the nuclear power plants (NPPs) and industrial installations in general. Due to historical reasons, the PSA sometimes is called PRA.

As the most important area of PSA projects remains nuclear power plants, mainly due to the specific features of the nuclear installations, three levels of PSA have evolved:

Level 1: The assessment of plant failures leading to core damage and the estimation of core damage frequency. A Level 1 PSA provides insights into design weaknesses and ways of preventing core damage. In the case of other industrial assessments, Level 1 PSA provides estimates of the accidents frequency and the main contributors.

Level 2: As possible releases are additionally protected by containment in most NPPs, PSA at this response and severe accident management possibilities. The results obtained in Level 1 are the basis for Level 2 quantification. In the case of other industrial assessments, Level 2 PSA might be fully covered by Level 1, as containment function is rather unique feature and is not common in other industries.

Level 3: The assessment of off-site consequences leading to estimates of risks to the public. Level 3 incorporates results om both previous levels.

Level1 PSA is the most important level and creates the background for further risk assessment, therefore it will be presented in detail. The structure of the other levels is much more application specific, and will be discussed only in general.

The methodology is based on systematically: 1) postulating potential accident scenarios triggered by an initiating event (IE), 2) identifying the systems acting as "defences" against these scenarios, 3) decomposing the systems into components, associating the failure modes and relative probabilities, 4) assessing the frequency of the accident scenarios. Two elements of the PSA methodology typically stand out:

- The event tree (ET) which is used to model the accident scenarios: it represents the main sequences of functional success and failure of safety systems appointed to cope with the initiating events and the consequences of each sequence. These consequences, denoted also as end states, are identified either as a safe end state or an accident end state.
- The fault tree (FT) which documents the systematic, deductive analysis of all the possible causes for the failure of the required function within an accident scenario modelled by the ET. A FT analysis is performed for each of the safety systems, required in response to the IE.

Assigning the safe end state to a sequence means that the scenario has been successfully terminated and undesired consequences have not occurred. In contrast the accident end state means that the sequence has resulted in undesired consequences.

Synthetically, the methodology embraced for the analysis consists of the following major tasks:

- identification of initiating events or initiating event groups of accident sequences: each initiator is defined by a frequency of occurrence;
- systems analysis: identification of functions to be performed in response to each initiating events to successfully prevent plant damage or to mitigate the consequences

and identification of the correspondent plant systems that perform these functions (termed front-line systems): for each system the probability of failure is assessed, by fault tree model;

- accident sequences development by constructing event trees for each initiating event or initiating event groups;
- accident sequences analysis to assess the frequencies of all relevant accident sequences;
- identification of dominant sequences on a frequency-consequence base, i.e. the ones presenting the most severe consequences to the personnel, the plant, the public and the environment and definition of the reference accident scenarios to be further analysed through deterministic transient analysis (for instance by t-h code simulation), in order to verify the fulfilment of the safety criteria. Consequences in the case of Level 1 PSA of NPPs are usually defined as degrees of reactor core damage, including 'safe' state and 'severe' accident state.

One of the main issues encountered in probabilistic analysis concerns the availability of pertinent data for the quantification of the risk, which eventually raises a large uncertainty in the results achieved. Usually these data are accessible from consolidated data bases (e.g. IAEA), resulting from the operational experience of the plants.

They pertain, for instance, to component failure rates, component probability on demand, initiating event frequency: for this reason within a PSA study usually an uncertainty analysis, in addition to a sensitivity analysis, is required in order to add credit to the model and to assess if sequences have been correctly evaluated on the probabilistic standpoint.

Event trees are used for the graphical and logical presentation of the accident sequences. An example of an event tree is shown in Figure 7. The logical combinations of success/failure conditions of functions or systems (usually safety systems, also called front-line systems) in the event tree are modelled by the fault tree.

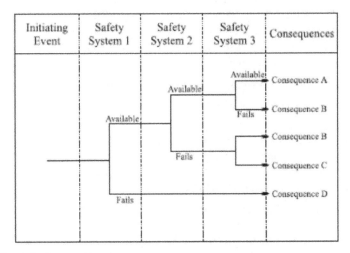

Figure 7. Example of an event tree

A fault tree logically combines the top event (e.g. complete failure of a support system) and the causes for that event (e.g. equipment failure, operator error etc.). An example of the fault tree is shown in Figure 8. The fault tree mainly consists of the basic events (all possible causes of the top event that are consistent with the level of detail of the study) and logical gates (OR, AND, M out of N and other logical operations). Other modelling tools, like common cause failures, house or area events are also used in the fault trees. All front-line and support systems are modelled by the fault trees and then combined in the event trees depending on the initiating event.

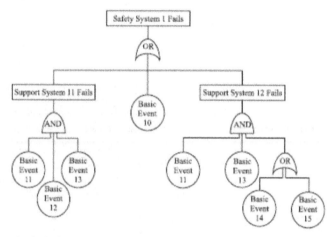

Figure 8. Example of a fault tree

A fault tree is capable to include rather special cases, usually identified in complex systems. These include system and components dependencies, called common cause failures (simultaneous failures of several components due to the same reason), area events (usually fire, flood etc., which damages groups of components in certain rooms), human actions (operator errors or mitigation actions).

The PSA is a powerful tool that can be used in many different ways to assess, understand and manage risk. Its primarily objectives are the following:

-   estimate risk level of the facility,
-   identify dominant event sequences affecting safety of the facility,
-   identify systems, components and human actions important for safety,
-   assess important dependencies (among systems or man-machine interactions),
-   provide decision support in various application areas.

The growing area of PSA use is extensive support of probabilistic results in risk management and decision-making processes. The main areas of the PSA applications are assessment of design modifications and back-fitting, risk informed optimization of the Technical Specifications, accident management, emergency planning and others. Several modern tools of risk management are also based on the PSA model, such as risk monitoring, precursor analysis and others.

Despite its popularity among the risk assessment tools, the PSA has a number of imitations and drawbacks. The main limitations of the PSA model are the following:

Binary representation of the component state. Only two states are analyzed: failed state or fully functioning state. However, this is not always realistic, as intermediate states are also possible. The same limitation exists for the redundant systems with certain success criteria - system is in failed state (success criteria is not satisfied) or in full power. The intermediate states for redundant systems are even more important.

Independence. In most cases, the components are assumed to be independent (except modelled by CCF), however there are many sources of dependencies, not treated by the model.

Aging effect. The aging effect is ignored because of the constant failure rate assumption. The only conservative possibility to treat the aging impact is to perform sensitivity study.

Time treatment. The FT/ET model is not capable to treat time explicitly during the accident progression. This is one of the major drawbacks of the methodology. In realistic systems, many parameters and functions depend on time and this is not encountered in the model and only approximate chronological order is assumed.

Uncertainty of the calculations. Uncertainties are inevitable in the PSA results and calculations and therefore direct treatment of the quantitative PSA estimates might be misleading. Due to the fact of uncertainties, the qualitative PSA results (identification of dominant accident sequences, comparison of different safety modifications) are of greater importance than quantitative.

4. Passive system unavailability model

The reliability of a passive system refers to the ability of the system to carry out a safety function under the prevailing conditions when required and addresses mainly the related performance stability.

In general the reliability of passive systems should be seen from two main aspects:

- systems/components reliability (e.g. piping, valves), as, for instance, the failure to start-up the system operation (e.g. drain valve failure to open)
- physical phenomena reliability, which addresses mainly the natural circulation stability, and the proneness of the system to the failure is dependent on the boundary conditions and the mechanisms needed for maintaining the intrinsic phenomena rather than on component malfunctions.

These two kinds of system malfunction are to to be considered as ET headings, to be assessed by specific FT components, as shown in figures 9 and 10.

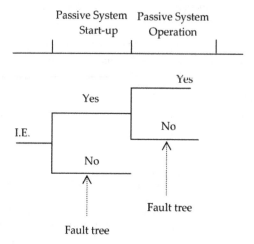

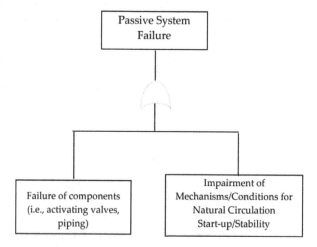

Figure 9. Event tree development

Figure 10. Fault tree model

The first facet calls for well-engineered safety components with at least the same level of reliability of the active ones.

The second aspect is concerned with the way the physical principle (gravity and density difference) operate and depends on the surrounding conditions related to accident development in terms of thermal hydraulic parameters evolution (i.e. characteristic parameters as flow rate and exchanged heat flux). This could require not a unique unreliability figure, but the unreliability to be re evaluated for each sequence following an accident initiator, or at least for a small group of bounding accident sequences, enveloping the ones chosen upon similarity of accident progress and expected consequences: with this respect thermal hydraulic analysis of the accident is helpful to estimate the evolution of the parameters during the accident progress.

First step of the analysis is the identification of the failure modes affecting the natural circulation: for this scope two well structured commonly used qualitative hazard analysis, as Failure Mode and Effect Analysis (FMEA) and HAZard and OPerability analysis (HAZOP), specifically tailored on the topic, by considering the phenomenology typical of natural circulation, are adopted.

This analysis concerns both mechanical components (e.g. valve, piping, heat exchanger) of the system and the natural circulation itself, as "virtual" component and the system under investigation is the aforementioned Isolation Condenser.

FMEA is a bottom-up procedure conducted at component level by which each failure mode in a system is investigated in terms of failure causes, preventive actions on causes, consequences on the system, corrective/preventive actions to mitigate the effects on the system, while the HAZOP procedure considers any parameters characteristic of the system (among pressure, temperature, flow rate, heat exchanged through the HX, opening of the drain valve) and by applying a set of "guide" words, which imply a deviation from the nominal conditions as for instance undesired decrease or increase, determines the consequences of operating conditions outside the design intentions. FMEA and HAZOP analysis are shown in Table 1 and 2 respectively.

The analysis points out several factors leading to disturbances in the Isolation Condenser system; the list of these includes:

- Unexpected mechanical and thermal loads, challenging the primary boundary integrity
- HX plugging
- Mechanical component malfunction, i.e. drain valve
- Non-condensable gas build-up
- Heat exchange process reduction: surface oxidation, thermal stratification, piping layout, etc.

Finally a set of critical parameters direct indicators of the failure of the system is identified; these include:

- Non-condensable fraction
- Undetected leakage
- Valve closure area in the discharge line
- Heat loss
- Piping layout
- HX plugged pipes

Component	Failure Mode	Causes	Prev. Actions on Causes	Consequences	Corrective/Preventive Action on Consequences	Comment
System piping	Rupture	Material defects and aging; Corrosion; Abnormal operation conditions; Vibrations; Local. Stresses; Impact of heavy loads (missile)	Adequate welding process quality; Water chemistry control; In Service inspect; Design against missile generation	LOCA in the Drywell; Instantaneous loss of natural circulation; Emptying of the circuit; Loss of heat removal capability; Loss of reactor coolant inventory	Isolate the breached loop; Safety relief valves actuation; Automatic reactor depressurisation; Gravity Driven Cooling System actuation;	Includes both steam line and drain line Critical Parameter: Undetected Leakage
	Leak	Material defects and aging; Corrosion; Abnormal operation conditions; Vibrations; Local. stresses	Adequate welding process quality; Water chemistry control; In Service inspect.	Small LOCA in the Drywell; Slow emptying of the circuit and natural circulation arrest for long periods of operation; Reduced heat removal capability	Leak monitoring; Isolate the breached loop; Safety relief valves actuation	Critical Parameter: Undetected Leakage
Tube Bundle of the heat exchangers of the IC	Single pipe rupture	Wearing due to vibration and corrosion	Preventive maintenance; Water chemistry control; Leak monitoring	Release of primary water to the pool; Slow emptying of the circuit and natural circulation arrest for long periods of operation; Reduced heat removal capability	Flow monitoring; Isolate the breached loop; Safety relief valves actuation	Critical Parameter: Undetected Leakage
	Multiple pipe rupture	Wearing due to corrosion, vibration and pressure transient	Preventive maintenance; Water chemistry control; Leak monitoring	Release of primary water to the pool; Natural circulation stop; Emptying of the circuit; Loss of reactor coolant inventory; Loss of heat removal capability	Isolate the breached loop; Safety relief valves actuation; Automatic reactor depressurisation; Gravity Driven Cooling System actuation	Critical Parameter: Undetected Leakage
	Single pipe plugging	Crud in the cooling loop; Foreign object in the cooling loop	Water chemistry control; Yearly test of pipes flow; Preventive maintenance	No consequences		Critical Parameter: HX Plugged Pipes
Tube Bundle of the heat exchangers of the IC	Multiple pipe plugging	Violent pressure and vibration transient detaching large amount of crud from pipes walls.	Water chemistry control; Use of suitable materials for cooling loop pipes; Preventive maintenance	Natural circulation stop; Loss of heat removal capability; Reactor pressure and temperature increase	Safety relief valves actuation	Critical Parameter: HX Plugged Pipes

Component	Failure Mode	Causes	Prev. Actions on Causes	Consequences	Corrective/Preventive Action on Consequences	Comment
Drain valve on the return condensate line	Valve fails to open	Control circuit failure; Loss of electric power to motor; Electric motor failure	Redundancy of control devices; Signal to the operator; In Service inspect.	Non triggering of Isolation Condenser if bypass valve does not operate; Loss of heat removal capability; Reactor pressure and temperature increase	Reactor pressure and temperature control; Safety relief valves actuation; Realignment by the operator; Corrective maintenance	Critical Parameter: Partially Open Valve
	Inadvertent valve closing	Spurious signal; Control circuit failure; Human error	Redundancy of control devices; Signal to the operator; Procedured actions	Natural circulation stop in case bypass valve does not operate; Loss of heat removal capability; Reactor pressure and temperature increase	Reactor pressure and temperature control; Safety relief valves actuation; Realignment by the operator; Corrective maintenance	Critical Parameter: Partially Open Valve
Natural Circulation	Envelope failure	Material defects and aging; Corrosion; Abnormal operation conditions; Vibrations; Local. Stresses; Impact of heavy loads (missile)	Adequate welding process quality; Water chemistry control; In Service inspect; Design against missile generation	LOCA in the Drywell; Instantaneous loss of natural circulation; Emptying of the circuit; Loss of heat removal capability; Loss of reactor coolant inventory	Isolate the breached loop; Safety relief valves actuation; Automatic reactor depressurisation; Gravity Driven Cooling System actuation	Includes both steam line and drain line Critical Parameter: Undetected Leakage
	Cracking	Material defects and aging; Corrosion; Abnormal operation conditions; Vibrations; Local. stresses	Adequate welding process quality; Water chemistry control; In Service inspect.	Small LOCA in the Drywell; Slow emptying of the circuit and natural circulation arrest for long periods of operation; Reduced heat removal capability	Leak monitoring; Isolate the breached loop; Safety relief valves actuation	Critical Parameter: Undetected Leakage
	Modification of surface characteristics	Oxidation; Aerosol deposits	Water chemistry control	Reduction in heat exchange efficiency; Reduced heat removal capability	Flow monitoring	Critical Parameter: Oxide Layer
Natural Circulation	Thermal stratification	Temperature dishomogeneity; Density variations; Onset of local thermal hydraulic phenomena	Process control (pressure, flow, temperature)	Reduction of heat convection; Natural circulation blockage; Loss of heat removal capability; Reactor pressure and temperature increase	Flow monitoring; Reactor pressure and temperature control; Safety relief valves actuation	Critical Parameter: Piping Layout, Heat Loss

Component	Failure Mode	Causes	Prev. Actions on Causes	Consequences	Corrective/Preventive Action on Consequences	Comment
	Non condensable build-up	Onset of chemical phenomena; Radiolysis products; Impurities	Water chemistry control (PH, O2, H2)	Reduction in heat exchange efficiency; Reduction of heat convection; Natural circulation blockage; Loss of heat removal capability; Reactor pressure and temperature increase	Flow monitoring; Reactor pressure and temperature control; Purging through vent lines Safety relief valves actuation	Critical Parameter: Non-Condensable Fraction
	Heat dissipation	Thermal insulation degradation; Inaccurate material assembly	In Service inspect.	Reduction of heat convection; Natural circulation impairment	Flow monitoring;	Critical Parameter: Heat Loss

Table 1. FMEA Table for the Isolation Condenser System

PARAMETER: Flow rate					
Guide Word	Deviation	Possible Causes	Consequences	Safeguards/Interlocks	Actions Required
More of[1]	High Flow	N/A			
Less of	Low Flow	Modifications of surface characteristics (crud deposition, oxidation); Non-condensable build-up; Thermal stratification; Pipe partial plugging; Pipe leak; HX single pipe plugging; HX single pipe rupture Drain valve partial opening	Natural circulation degradation and reduced heat transfer capability; T and P increase	Safety relief valve actuation; Vent line valve actuation	Corrective maintenance; Operator action
No/None	No Flow	Non-condensable build-up; Thermal stratification; Pipe plugging; Pipe rupture; HX Multiple pipe plugging; HX Multiple pipe rupture; Drain valve closed	Natural circulation stop and loss of heat transfer capability; T and P increase	Safety relief valve actuation; Vent line valve actuation; Automatic Depressurisation System actuation; Gravity Driven Cooling System actuation	Corrective maintenance; Operator action
PARAMETER: Pressure					
Guide Word	Deviation	Possible Causes	Consequences	Safeguards/Interlocks	Actions Required
More of	High Pressure	Non-condensable build-up; Surface modifications (crud, oxidation); HX tube plugging; HX tube rupture Partial valve opening	Natural circulation degradation and reduced heat transfer capability; T increase	Safety relief valve actuation; Vent line valve actuation	Corrective maintenance; Operator action
Less of[1]	Low Pressure	N/A			
No/None	No Pressure	N/A			

PARAMETER: Drain valve opening					
Guide Word	Deviation	Possible Causes	Consequences	Safeguards/Interlocks	Actions Required
More of	N/A				
Less of	Reduced Opening	Partial blockage	Natural circulation degradation and reduced heat transfer capability; T and P increase	Safety relief valve actuation;	Corrective maintenance; Operator action
No/None	No Opening	Loss of electrical power; Circuit control failure; Electrical motor failure; Valve stuck	Natural circulation stop and loss of heat transfer capability; T and P increase	Safety relief valve actuation; Vent line valve actuation; Automatic Depressurisation System actuation; Gravity Driven Cooling System actuation	Corrective maintenance; Operator action

PARAMETER: Exchanged heat flux					
Guide Word	Deviation	Possible Causes	Consequences	Safeguards/Interlocks	Actions Required
More of[1]	High flux	N/A			
Less of	Low Flux	Non-condensable build-up; Surface modifications (crud, oxidation); HX single tube plugging; HX single tube rupture	Natural circulation degradation and reduced heat transfer capability; T and P increase	Safety relief valve actuation; Vent line valve actuation	Corrective maintenance; Operator action
No/None	No Flux	Non-condensable build-up; HX multiple tube plugging; HX multiple tube rupture	Natural circulation stop and loss of heat transfer capability; T and P increase	Safety relief valve actuation; Vent line valve actuation; Automatic Depressurisation System actuation; Gravity Driven Cooling System actuation	Corrective maintenance; Operator action

[1] This deviation is not evaluated, even if it implies an overcooling of the system that could potentially induce to thermal stresses on core structures and reactor components, like the heat exchanger.

Table 2. HAZOP Table for the Isolation Condenser System

Each of these failure mode driving parameters is examined to determine the expected failure probability by defining the range and the probability distribution function pertaining to the parameter. These failure characteristics are then used to develop a probabilistic model to predict the natural circulation failure.

As stated before FT technique seems to be the most suitable mean to quantify the passive system unavailability, once introduced the failure modes in the form of critical parameters elementary basic events, linked following the Boolean algebra rules (AND et OR), or in the form of sub-fault trees. However the introduction of passive safety systems into an accident scenario, in the fashion of a safety or front line system, deserves particular attention. The reason is that its reliability figure depends more on the phenomenological nature of occurrence of the failure modes rather than on the classical component mechanical and electrical faults. This makes the relative assessment process different as regards the system model commonly adopted in the fault tree approach as depicted before.

In fact, since the failure of the physical process is addressed, the conventional failure model associated with the basic events (i.e. exponential, $e^{-\lambda t}$, λ failure rate, t mission time), commonly used for component failure model, is not applicable: each pertinent basic event will be characterized by defined parameters driving the failure mechanisms - e.g. non-condensable fraction, leak rate, partial opening of the isolation valve, heat exchanger plugged pipes, etc. - and the associated failure criterion. Thus each basic event model pertaining to the relevant failure mode requires the assignment of both the probability distribution and range of the correspondent parameter and the definition of the critical interval defining the failure (for example failure for non-condensable fraction >x%, leak rate > x gr./sec or crack size > x cm^2 and so on).In order to evaluate the overall probability of failure of the system, the single failure probabilities are combined according to:

$$Pe_t = 1.0- ((1.0 - Pe_1)*(1.0 - Pe_2)*...*(1.0 - Pe_n)) \qquad (1)$$

where:

Pe_t overall probability of failure

Pe_1 through Pe_n individual probabilities of failure pertaining to each failure mode, assuming mutually non-exclusive independent events

The failure model relative to each single basic event is given by:

$$Pe_i= \int p_i(x) \, dx \quad x>x_0 \qquad (2)$$

$p_i(x)$ probability distribution function of the parameter x

x_0 threshold value according to the failure criterion

It's worth noting that the assumed failure criterion, based on the failure threshold for each path, implies the neglecting of the "intermediate" modes of operation of the system or equivalently the degraded performance of the system (up to the failure point): this gives credit for a passive system that "partially works" and has failed for its intended function but provides some operation. This operation could be sufficient to prolong the window for opportunity to recover a failed system, for instance through redundancy configuration, and ultimately prevent or arrest core degradation.

Once the probabilistic distributions of the parameters are assigned, the reliability of the system can be directly obtained from (1) once a failure criterion is assigned and the single failure probabilities are evaluated through (2): this point is being satisfied by assigning both the range and the probability distributions, basing on expert judgment and engineering assessment. In fact, as further illustrated, difficulties arise in assigning both the range and the probability density functions relative to the critical parameters defining the failure modes, in addition to the definition of a proper failure criterion, because of the lack of operational experience and data.

5. Methodologies characterization and comparative assessment

A very good description of the various methodologies proposed so far and currently available in the open literature is given in [5].

The earliest significant effort to quantify the reliability of such systems is represented by a methodology known as REPAS (Reliability Evaluation of Passive Systems), [6], which has been developed in late 1990s, cooperatively by ENEA, the University of Pisa, the Polytechnic of Milan and the University of Rome, that was later incorporated in the EU (European Union) RMPS (Reliability Methods for Passive Systems) project. This methodology is based on the evaluation of a failure probability of a system to carry out the desired function from the epistemic uncertainties of those physical and geometric parameters which can cause a failure of the system.

The RMPS methodology, described in [7], was developed to address the following problems: 1) Identification and quantification of the sources of uncertainties and determination of the important variables, 2) Propagation of the uncertainties through thermal-hydraulic (T-H) models and assessment of passive system unreliability and 3) Introduction of passive system unreliability in accident sequence analyses. In this approach, the passive system is modelled by a qualified T-H code (e.g. CATHARE, RELAP) and the reliability evaluation is based on results of code runs, whose inputs are sampled by Monte-Carlo (M-C) simulation. This approach provides realistic assessment of the passive system reliability, thanks to the flexibility of the M-C simulation, which adapts to T-H model complexity without resort to simplifying approximation. In order to limit the number of T-H code runs required by M-C simulation, alternative methods have been proposed such as variance reduction techniques, first and second order reliability methods and response surface methods. The RMPS methodology has been successfully applied to passive systems utilizing natural circulation in different types of reactors (BWR, PWR, and VVER). A complete example of application concerning the passive residual heat removal system of a CAREM reactor is presented in [8]. The RMPS methodology tackles also an important problem, which is the integration of passive system reliability in a PSA study. So far, in existing innovative nuclear reactor projects PSA's, only passive system components failure probabilities are taken into account, disregarding the physical phenomena on which the system is based, such as the natural circulation. The first attempts performed within the framework of RMPS have taken into account the failures of the components of the passive system as well as the impairment of the physical process involved like basic events in static event tree as exposed in [7]. Two other steps have been identified after the development of the RMPS methodology where an improvement was desirable: the inclusion of a formal expert judgment (EJ) protocol to estimate distributions for parameters whose values are either sparse on not available, and the use of efficient sensitivity analysis techniques to estimate the impact of changes in the input parameter distributions on the reliability estimates.

R&D in the United States on the reliability of passive safety systems has not been as active at least until mid 2000. A few published papers from the Massachusetts Institute of Technology (MIT) have demonstrated their development of approaches to the issue. Their technique has examined TH uncertainties in passive cooling systems for Generation IV-type gas-cooled reactors. The MIT research on the reliability of passive safety systems has taken a similar approach but has focused on a different set of reactor technologies. Their research has examined thermal hydraulic uncertainties in passive cooling systems for Generation IV gas-cooled reactors, as described in [9,10]. Instead of post-design probabilistic risk analysis for regulatory purposes, the MIT research seeks to leverage the capabilities of probabilistic risk assessment (PRA) to improve the design of the reactor systems early in their development life cycle.

In addition to the RMPS approach, a number of alternative methodologies have been investigated for the reliability assessment of T H passive systems.

Three different methodologies have been proposed by ENEA (Italian National Agency for New Technologies, Energy and Sustainable Economic Development). In the first methodology [11], the failure probability is evaluated as the probability of occurrence of different independent failure modes, a priori identified as leading to the violation of the boundary conditions or physical mechanisms needed for successful passive system operation.

This approach based on independent failure modes introduces a high level of conservatism as it appears that the probability of failure of the system is relevantly high, because of the combination of various modes of failure as in a series system, where a single fault is sufficient to challenge the system performance. The correspondent value of probability of failure can be conservatively assumed as the upper bound for the unavailability of the system, within a sort of "parts-count" reliability estimation.

In the second, [12], modelling of the passive system is simplified by linking to the modelling of the unreliability of the hardware components of the system: this is achieved by identifying the hardware failures that degrade the natural mechanisms upon which the passive system relies and associating the unreliability of the components designed to assure the best conditions for passive function performance.

Thus, the probabilities of degraded physical mechanisms are reduced to unreliability figures of the components whose failures challenge the successful passive system operation. If, on the one hand, this approach may in theory represent a viable way to address the matter, on the other hand, some critical issues arise with respect to the effectiveness and completeness of the performance assessment over the entire range of possible failure modes that the system may potentially undergo and their association to corresponding hardware failures. In this simplified methodology, degradation of the natural circulation process is always related to failures of active and passive components, not acknowledging, for instance, any possibility of failure just because of unfavourable initial or boundary conditions. In addition, the fault tree model adopted to represent the physical process decomposition is used as a surrogate model to replace the complex T-H code that models the system behaviour. This decomposition is not appropriate to predict interactions among physical phenomena and makes it extremely difficult to realistically assess the impact of parametric uncertainty on the performance of the system.

The third approach is based on the concept of functional failure, within the reliability physics framework of load-capacity exceedance [7,13,14]. The functional reliability concept is defined as the probability of the passive system failing to achieve its safety function as specified in terms of a given safety variable crossing a fixed safety threshold, leading the load imposed on the system to overcome its capacity. In this framework, probability distributions are assigned to both safety functional requirement on a safety physical parameter (for example, a minimum threshold value of water mass flow required to be circulating through the system for its successful performance) and system state (i.e., the actual value of water mass flow circulating), to reflect the uncertainties in both the safety thresholds for failure and the actual conditions of the system state. Thus the mission of the passive system defines which parameter values are considered a failure by comparing the corresponding pdfs according to defined safety criteria. The main drawback in the last method devised by ENEA lies in the selection and definition of the probability distributions that describe the characteristic parameters, based mainly on subjective/engineering judgment.

Every one of three methods devised by ENEA shares with the main RMPS approach the issue related to the uncertainties affecting the system performance assessment process. With respect to the RMPS a greater simplicity is introduced, although detrimental to the relevance of the approaches themselves: this is particularly relevant as far as the approach based on hardware components failure is concerned.

Finally a different approach is followed in the APSRA (Assessment of Passive System ReliAbility) methodology developed by BARC (Bhabha Atomic Research Centre, India), see [15]. In this approach, a failure surface is generated by considering the deviation of all those critical parameters, which influence the system performance. Then, the causes of deviation of these parameters are found through root diagnosis. It is attributed that the deviation of such physical parameters occurs only due to a failure of mechanical components such as valves, control systems, etc. Then, the probability of failure of a system is evaluated from the failure probability of these mechanical components through classical PSA treatment. Moreover, to reduce the uncertainty in code predictions, BARC foresee to use in-house experimental data from integral facilities as well as separate.

With reference to the two most relevant methodologies (i.e. RMPS and APSRA), the RMPS consists mainly in the identification and quantification of parameter uncertainties in the form of probability distributions, to be propagated directly into a T-H code or indirectly in using a response surface; the APSRA methodology strives to assess not the uncertainty of parameters but the causes of deviation from nominal conditions, which can be in the failure of active or passive components or systems.

As a result, different approaches are used in the RMPS and APSRA methodologies. RMPS proposes to take into account, in the PSA model, the failure of a physical process. This problem is treated in using a best estimate T-H code plus uncertainty approach. APSRA includes in the PSA model the failure of those components which cause a deviation of the key parameters resulting in a system failure, but does not take into account possible uncertainties on these key parameters. As the consequence, the T-H code is used in RMPS to propagate the uncertainties and in APSRA to build a failure surface. APSRA incorporates an important effort on qualification of the model and use of the available experimental data. These aspects have not been studied in the RMPS, given the context of the RMPS project.

The following Table attempts to identify the main characteristics of the methodologies proposed so far, with respect to some aspects, such as the development of deterministic and probabilistic approaches, the use of deterministic models to evaluate the system performance, the identification of the sources of uncertainties and the application of expert judgment.

Methodology	Probabilistic vs. deterministic	Deterministic Analysis	Uncertainties	Expert Judgment/Experimental data
REPAS/RMPS	Merge of probabilistic and thermal hydraulic aspects	T-H code adopted for uncertainty propagation	Uncertainties in parameters modelled by probability density functions	EJ adopted to a large extent; Statistical analysis when experimental data exist
APSRA	Merge of probabilistic and thermal hydraulic aspects	T-H code adopted to build the failure surface	parameters' deviations from nominal conditions caused by failure of active or passive components (root diagnosis)	Experimental data usage; EJ for root diagnosis
ENEA approaches	Only probabilistic aspects		Uncertainties in parameters	EJ adopted to a large extent (except the approach based on hardware failure)

Table 3. Main features of the various approaches

6. Open issues

From the exam of the various methodologies, which have been developed over these most recent years within the community of the safety research, and are currently available in the open literature, the following open questions are highlighted and consequently needs for research in all related areas are pointed out :

- The aspects relative to the assessment of the uncertainties related to passive system performance: they regard both the best estimate T-H codes used for their evaluation and system reliability assessment itself;
- The dependencies among the parameters, mostly T-H parameters, playing a key role in the whole process assessment.
- The integration of the passive systems within an accident sequence in combination with active systems and human actions.
- The consideration for the physical process and involved physical quantities dependence upon time, implying, for instance, the development of dynamic event tree to incorporate the interactions between the physical parameter evolution and the state of the system and/or the transition of the system from one state to another.

It's worth noticing that these two last aspects are correlated, but hey will be treated separately.

- The comparison between active and passive systems, mainly on a functional viewpoint.

All of these points are elaborated in the following, in an attempt to cover the entire spectrum of issues related to the topic, and capture all the relevant aspects to concentrate on and devote resources towards for fulfilling a significant advance.

6.1. Uncertainties

The quantity of uncertainties affecting the operation of the T-H passive systems affects considerably the relative process devoted to reliability evaluation, within a probabilistic safety analysis framework, as recognized in [7].

These uncertainties stem mainly from the deviations of the natural forces or physical principles, upon which they rely (e.g., gravity and density difference), from the expected conditions due to the inception of T-H factors impairing the system performance or to changes of the initial and boundary conditions, so that the passive system may fail to meet the required function. Indeed a lot of uncertainties arise, when addressing these phenomena, most of them being almost unknown due mainly to the scarcity of operational and experimental data and, consequently, difficulties arise in performing meaningful reliability analysis and deriving credible reliability figures. This is usually designated as phenomenological uncertainty, which becomes particularly relevant when innovative or untested technologies are applied, eventually contributing significantly to the overall uncertainty related to the reliability assessment.

Actually there are two facets to this uncertainty, i.e., "aleatory" and "epistemic" that, because of their natures, must be treated differently. The aleatory uncertainty is that addressed when the phenomena or events being modelled are characterized as occurring in a "random" or "stochastic" manner and probabilistic models are adopted to describe their occurrences. The epistemic uncertainty is that associated with the analyst's confidence in the prediction of the PSA model itself, and it reflects the analyst's assessment of how well the PSA model represents the actual system to be modelled. This has also been referred to as state-of-knowledge uncertainty, which is suitable to reduction as opposed to the aleatory which is, by its nature, irreducible. The uncertainties concerned with the reliability of passive system are both stochastic, because of the randomness of phenomena occurrence, and of epistemic nature, i.e. related to the state of knowledge about the phenomena, because of the lack of significant operational and experimental data.

For instance, as initial step, the approach described in [16]. allows identifying the uncertainties pertaining to passive system operation in terms of critical parameters driving the modes of failure, as, for instance, the presence of non-condensable gas, thermal stratification and so on. In this context the critical parameters are recognized as epistemic uncertainties.

The same reference points out, as well, the difference between the uncertainties related to passive system reliability and the uncertainties related to the T-H codes (e.g. RELAP), utilized to evaluate the performance itself, as the ones related to the coefficients, correlations, nodalization, etc.: these specific uncertainties, of epistemic nature, in turn affect the overall uncertainty in T-H passive system performance and impinge on the final sought reliability figure.

A further step of the matter can be found in[11], which attempts to assign sound distributions to the critical parameters, to further develop a probabilistic model. As is of common use when the availability of data is limited, subjective probability distributions are elicited from expert/engineering judgment procedure, to characterize the critical parameters.

Three following classes of uncertainties to be addressed are identified:

- Geometrical properties: this category of uncertainty is generally concerned with the variations between the as-built system layout and the design utilized in the analysis: this is very relevant for the piping layout (e.g. suction pipe inclination at the inlet of the heat exchanger, in the isolation condenser reference configuration) and heat loss modes of failure.
- Material properties: material properties are very important in estimating the failure modes concerning for instance the undetected leakages and the heat loss.
- Design parameters, corresponding to the initial/boundary conditions (for instance, the actual values taken by design parameters, like the pressure in the reactor pressure vessel).
- Phenomenological analysis: the natural circulation failure assessment is very sensitive to uncertainties in parameters and models used in the thermal hydraulic analysis of the system. Some of the sources of uncertainties include but are not limited to: the definition of failure of the system used in the analysis, the simplified model used in the analysis, the analysis method and the analysis focus on failure locations and modes and finally the selection of the parameters affecting the system performance.

The first, second and third groups are part of the category of aleatory uncertainties because they represent the stochastic variability of the analysis inputs and they are not reducible.

The fourth category is referred to the epistemic uncertainties, due to the lack of knowledge about the observed phenomenon and thus suitable for reduction by gathering a relevant amount of information and data. This class of uncertainties must be subjectively evaluated, since no complete investigation of these uncertainties is available.

A clear prospect of the uncertainties as shown in Table 4 [5].

As emphasized above, clearly the epistemic uncertainties address mostly the phenomena underlying the passive operation and the parameters and models used in the T-H analysis of the system (including the ones related to the best estimate code) and the system failure analysis itself. Some of the sources of uncertainties include but are not limited to the definition of failure of the system used in the analysis, the simplified model used in the analysis, the analysis method and the analysis focus of failure locations and modes and finally the selection of the parameters affecting the system performance. With this respect, it is important to underline, again, that the lack of relevant reliability and operational data imposes the reliance on the underlying expert judgment for an adequate treatment of the uncertainties, thus making the results conditional upon the expert judgment elicitation process. This can range from the simple engineering/subjective assessment to a well structured procedure based on expert judgment elicitation, as reported in [17], which outlines the main aspects of the REPAS procedure.

Aleatory
Geometrical properties
Material properties
Initial/boundary conditions (design parameters)
Epistemic
T-H analysis
Model (correlations)
Parameters
System failure analysis
Failure criteria
Failure modes (critical parameters)

Table 4. Categories of uncertainties associated with T-H passive systems reliability assessment

In ref. [17], in order to simplify both the identification of the ranges and their corresponding probabilities, initially discrete values have been selected. As a general rule, a central pivot has been identified, and then the range has been extended to higher and lower values, if applicable. The pivot value represents the nominal condition for the parameter. The limits have been chosen in order to exclude unrealistic values or those values representing a limit zone for the operation demand of the passive system. Once the discrete ranges have been set up, discrete probability distributions have been associated, to represent the probabilities of occurrence of the values. As in the previous step, the general rule adopted is that the higher probability of occurrence corresponds to the nominal value for the parameter. Then lower probabilities have been assigned to the other values, as much low the probability as much wide the distance from the nominal value, as in a sort of Gaussian distribution.

Ultimately, as underlined in the previous section, the methodologies proposed in RMPS and within the studies conducted by MIT address the question by propagating the parameter and model uncertainties, by performing Monte Carlo simulations on the detailed T-H model based on a mechanistic code, and calculating the distribution of the safety variable and thus the probability of observing a value above the defined limit, according to the safety criterion.

6.2. Dependencies

Alike some other types of analyses for nuclear power plants, the documented experience with PSS reliability seems to focus on the analysis of one passive attribute at a time. In many cases, this may be sufficient, but for some advanced designs with multiple passive features, modelling of the synergistic effects among them is important. For example, modelling of a passive core cooling system may require simultaneous modelling of the amount of non condensable gases which build up along the circuit during extended periods of operation, the potential for stratification in the cooling pool, and interactions between the passive core cooling system and the core. Analysis of each of these aspects independently may not fully capture the important boundary conditions of each system. For instance, with regard to the aforementioned methodologies, the basic simplifying assumption of independence among system performance relevant parameters, as the degradation measures, means that the correlation among the critical parameter distributions is zero or is very low to be judged significant, so that the assessment of the failure probability is quite straightforward. If parameters have contributors to their uncertainty in common, the respective states of knowledge are dependent. As a consequence of this dependence, parameter values cannot be combined freely and independently. Instances of such limitations need to be identified and the dependencies need to be quantified. If the analyst knows of dependencies between parameters explicitly, multivariate distributions or conditional subjective pdfs (probability density functions) may be used. The dependence between the parameters can be also introduced by covariance matrices or by functional relations between the parameters.

As observed in [15], both REPAS and RMPS approaches adopt a probability density function (pdf) to treat variations of the critical parameters considered in the predictions of codes. To apply the methodology, one needs to have the pdf values of these parameters. However, it is difficult to assign accurate pdf treatment of these parameters, which ultimately define the functional failure, due to the scarcity of available data, both on an experimental and operational ground. Moreover, these parameters are not really independent ones to have deviation of their own. Rather deviations of them from their nominal conditions occur due to failure/malfunctioning of other components or as a result of the combination with different concomitant mechanisms. Thus the hypothesis of independence among the failure driving parameters appears non proper.

With reference to the functional reliability approach set forth in [13], the selected representative parameters defining the system performance, for instance coolant flow or exchanged thermal power, are properly modelled through the construction of joint probability functions in order to assess the correspondent functional reliability. A recent study shows how the assumption of independence between the marginal distributions to construct the joint probability distributions to evaluate system reliability adds conservatism to the analysis, [18]: for this reason the model is implemented to incorporate the correlations between the parameters, in the form of bivariate normal probability distributions. That study has the merit to highlight the dependence among the parameters underlying the system performance: further studies are underway, with regard, for instance to the approach based on independent failure modes. As described in the previous section 2, this approach begins by identifying critical parameters, properly modelled through probability functions, as input to basic events, corresponding to the failure modes, arranged in a series system configuration, assuming non-mutually exclusive independent events. It introduces a high level of conservatism as it appears that the probability of failure of the system is relevantly high to be considered acceptable, because of the combination of various modes of failure, where a single fault is sufficient to challenge the system performance. Initial evaluations, [19], reveal that the critical parameters are not suitable to be chosen independently of each other, mainly because of the expected synergism between the different phenomena under investigation, with the potential to jeopardize the system performance. This conclusion allows the implementation of the proposed methodology, by properly capturing the interaction between various failure modes, through modelling system performance under multiple degradation measures. It was verified that when the multiple degradation measures in a system are correlated, an incorrect independence assumption may overestimate the system reliability, according to a recent study, [20].

6.3. Incorporation of passive system within probabilistic safety assessment

PSA has been introduced for the evaluation of design and safety in the development of those reactors. A technology-neutral framework, that adopts PSA information as a major evaluation tool, has been proposed as the framework for the evaluation of safety or regulation for those reactors [21,22]. To utilize this framework, the evaluation of the reliability of Passive Systems has been recognized as an essential part of PSA.

In PSA, the status of individual systems such as a passive system is assessed by an accident sequence analysis to identify the integrated behaviour of a nuclear system and to assign its integrated system status, i.e. the end states of accident sequences. Because of the features specific of a passive system, it is difficult to define the status of a passive system in the accident sequence analysis. In other words, the status of a passive system does not become a robust form such as success or failure, since "intermediate" modes of operation of the system or equivalently the degraded performance of the system (up to the failure point) is possible. This gives credit for a passive system that "partially works" and has failed for its intended function but provides some operation: this operation could be sufficient to prolong the window for opportunity to recover a failed system, for instance through redundancy configuration, and ultimately prevent or arrest core degradation [19]. This means that the status of a passive system can be divided into several states, and each status is affected by the integrated behaviour of the reactor, because its individual performance is closely related with the accident evolution and whole plant behaviour.

Ref. [23] lays the foundations to outline a general approach for the integration of a passive system, in the form of a front line system and in combination with active ones and/or human actions, within a PSA framework.

In [7] a consistent approach, based on an event tree representation, has been developed to incorporate in a PSA study the results of reliability analyses of passive systems obtained on specific accident sequences. In this approach, the accident sequences are analyzed by taking into account the success or the failure of the components and of the physical process involved in the passive systems. This methodology allows the probabilistic evaluation of the influence of a passive system on a definite accident scenario and could be used to test the advantage of replacing an active system by a passive system in specific situations.

However in order to generalize the methodology, it is important to take into account the dynamic aspects differently than by their alone modelling into the T-H code. Indeed in complex situations where several safety systems are competing and where the human operation cannot be completely eliminated, this modelling should prove to be impossible or too expensive in computing times. It is thus interesting to explore other solutions already used in the dynamic PSA, like the method of the dynamic event trees, in order to capture the interaction between the process parameters and the system state within the dynamical evolution of the accident.

In the PSA of nuclear power plants (NPPs), accident scenarios, which are dynamic in nature, are usually analyzed with event trees and fault trees.

The current PSA framework has some limitations in handling the actual timing of events, whose variability may influence the successive evolution of the scenarios, and in modelling the interactions between the physical evolution of the process variables (temperatures, pressures, mass flows, etc.,) and the behaviour of the hardware components. Thus, differences in the sequential order of the same success and failure events and the timing of event occurrence along an accident scenario may affect its evolution and outcome; also, the evolution of the process variables (temperatures, pressures, mass flows, etc.,) may affect the event occurrence probabilities and thus the developing scenario. Another limitation lies in the binary representations of system states (i.e., success or failure), disregarding the intermediate states, which conversely concern the passive system operation, as illustrated above.

To overcome the above-mentioned limitations, dynamic methodologies have been investigated which attempt to capture the integrated response of the systems/components during an accident scenario [24].

The most evident difference between dynamic event trees (DETs) and the event trees (ETs) is as follows. ETs, which are typically used in the industrial PSA, are constructed by an analyst, and their branches are based on success/ failure criteria set by the analyst. These criteria are based on simulations of the plant dynamics. On the contrary, DETs are produced by a software that embeds the models that simulates the plant dynamics into stochastic models of components failure. A challenge arising from the dynamic approach to PSA is that the number of scenarios to be analyzed is much larger than that of the classical fault/event tree approaches, so that the a posteriori information retrieval can become quite onerous and complex.

This is even more relevant as far as thermal hydraulic natural circulation passive systems are concerned since their operation is strongly dependent, more than other safety systems, upon time and the state/parameter evolution of the system during the accident progression.

Merging probabilistic models with T-H models, i.e. dynamic reliability, is required to accomplish the evaluation process of T-H passive systems in a consistent manner: this is particularly relevant with regard to the introduction of a passive system in an accident sequence, since the required mission could be longer than 24 h as usual level 1 PSA mission time. In fact for design basis accidents, the passive systems are required to establish and maintain core cooling and containment integrity, with no operator intervention or requirement for a.c. power for 72 h, as a grace time [25].

The goal of dynamic PRA is to account for the interaction of the process dynamics and the stochastic nature/behavior of the system at various stages: it associates the state/parameter evaluation capability of the thermal hydraulic analysis to the dynamic event tree generation capability approach. The methodology should estimate the physical variation of all technical parameters and the frequency of the accident sequences when the dynamic effects are considered. If the component failure probabilities (e.g. valve per-demand probability) are known, then these probabilities can be combined with the probability distributions of estimated parameters in order to predict the probabilistic evolution of each scenario outcome.

A preliminary attempt in addressing the dynamic aspect of the system performance in the frame of passive system reliability is shown in [26], which introduces the T-H passive system as a non-stationary stochastic process, where the natural circulation is modeled in terms of time-variant performance parameters, (as for instance mass flow-rate and thermal power, to cite any) assumed as stochastic variables. In that work, the statistics associated with the stochastic variables change in time (in terms of associated mean values and standard deviations increase or decrease, for instance), so that the random variables have different values in every realization, and hence every realization is different.

6.4. Comparative assessment between active and passive systems

The design and development of future water-cooled reactors address the use of passive safety systems, i.e. those characterized by no or very limited reliance on external input (forces, power or signal, or human action) and whose operation takes advantage of natural forces, such as free convection and gravity, to fulfil the required safety function and to provide confidence in the plant's ability to handle transients and accidents. Therefore, they are required to accomplish their mission with a sufficient reliability margin that makes them attractive as an important means of achieving both simplification and cost reduction for future plants while assuring safety requirements with lesser dependence of the safety function on active components like pumps and diesel generators.

On the other hand, since the magnitude of the natural forces, which drive the operation of passive systems, is relatively small, counter-forces (e.g. friction) can be of comparable magnitude and cannot be ignored as is generally the case with pumped systems. This concern leads to the consideration that, despite the fact that passive systems "should be" or, at least, are considered, more reliable than active ones - because of the smaller unavailability due to hardware failure and human error - there is always a nonzero likelihood of the occurrence of physical phenomena leading to pertinent failure modes, once the system enters into operation.

These characteristics of a high level of uncertainty and low driving forces for heat removal purposes justify the comparative evaluation between passive and active options, with respect to the accomplishment of a defined safety function (e.g. decay heat removal) and the generally accepted viewpoint that passive system design is more reliable and more economical than active system design has to be discussed [27].

Here are some of the benefits and disadvantages of the passive systems that should be evaluated vs. the correspondent active system.

- Advantages
- No external power supply: no loss of power accident has to be considered.
- No human factor, implying no inclusion of the operator error in the analysis.
- Better impact on public acceptance, due to the presence of "natural forces".
- Less complex system than active and therefore economic competitiveness.
- Passive systems must be designed with consideration for ease of ISI, testing and maintenance so that the dose to the worker is much less.
- Drawbacks
- Reliance on "low driving forces", as a source of uncertainty, and therefore need for T-H uncertainties modeling.
- Licensing requirement (open issue), since the reliability has to be incorporated within the licensing process of the reactor. For instance the PRA's should be reviewed to determine the level of uncertainty included in the models.
- Need for operational tests, so that dependence upon human factor can not be neglected.
- Time response: the promptness of the system intervention is relevant to the safety function accomplishment. It appears that the inception of the passive system operation, as the natural circulation, is conditional upon the actuation of some active components (as the return valve opening) and the onset of the conditions/mechanisms for natural circulation start-up
- Reliability and performance assessment in any case. Quantification of their functional reliability from normal power operation to transients including accidental conditions needs to be evaluated. Functional failure can happen if the boundary conditions deviate from the specified value on which the performance of the system depends.
- Ageing of passive systems must be considered for longer plant life; for example corrosion and deposits on heat exchanger surfaces could impair their function.
- Economics of advanced reactors with passive systems, although claimed to be cheaper, must be estimated especially for construction and decommissioning.

The question whether it is favourable to adopt passive systems in the design of a new reactor to accomplish safety functions is still to be debated and a common consensus has not yet been reached, about the quantification of safety and cost benefits which make nuclear power more competitive, from potential annual maintenance cost reductions to safety system response.

7. Final remarks

Based on the analysis of the critical aspects related to the open points discussed in the previous section a qualitative analysis, on the basis of the author's opinion, reported in

Table 5 below aims at identifying for each of the above items both the criticality with respect to the passive system reliability assessment process, in terms of the relative importance and the existing advancement, according to Table 6 which ranks the relative level of both the importance and progress.

Item	Importance	Advance
Uncertainties	H	L
Dependencies	M	L
Integration within PSA	M	L
Passive vs. Active	H	L

Table 5. Importance analysis

	Grade	Definition
Importance	H	The item is expected to have a significant impact on the system failure
	M	The item is expected to have a moderate impact on the system failure
	L	The item is expected to have only a small impact on the system failure
Advance	H	The issue is modelled in a detailed way with adequate validation
	M	The issue is represented by simple modelling based on experimental observations or results.
	L	The issue is not represented in the analysis or the models are too complex or inappropriate which indicates that the calculation results will have a high degree of ambiguity

Table 6. Grade rank for importance and advancement analysis

It is clear that he worst case is characterized by "high "and "low" rankings relative respectively to the importance and the advancement aspects, thus making the correspondent item development a critical challenge.

Based on this, the results of this qualitative analysis show the relevance relative to the uncertainties and the comparison between active and passive, as most critical points to be addressed in the application of the PRA to the evaluation of the passive system performance assessment. This allows the analyst to track a viable R&D program to deal with these issues and limitations and to steer the relative efforts towards their implementation.

8. Conclusions

Due to the specificities of passive systems that utilize natural circulation (small driving force, large uncertainties in their performance, lack of data…), there is a strong need for the development and demonstration of consistent methodologies and approaches for evaluating their reliability. This is a crucial issue to be resolved for their extensive use in future nuclear power plants. Recently, the development of procedures suitable for establishing the performance of a passive system has been proposed: the unavailability of reference data makes troublesome the qualification of the achieved results. These procedures can be applied for evaluating the acceptability of a passive system, specifically when nuclear reactor safety considerations are important for comparing two different systems having the same mission and, with additional investigation, for evaluating the performance of an active and passive system on a common basis. The study while identifying limitations of the achieved results or specific significant aspects that have been overlooked has suggested areas for further development or improvements of the procedures:

- In order to get confidence in the achieved results, the reduction of the so identified level of uncertainty pertaining to the passive system behaviour, and regarding in particular the phenomenological uncertainty. In fact, it's worth noting that these uncertainties are mainly related to the state of knowledge about the studied object/phenomenon, i.e., they fall within the class of epistemic uncertainties, thus suitable for reduction by gathering and analyzing a relevant quantity of information and data.
- The determination of the dependencies among the relevant parameters adopted to analyze the system reliability.
- The study of the dynamical aspects of the system performance, because the inherent dynamic behaviour of the system to be characterized: this translates into the development of the dynamic event tree.
- The comparison against the active system, also to evaluate the economical competitiveness, while assuring the same level of safety.

Future research in nuclear safety addressing this specific topic relevant to advanced reactors should be steered towards all these points in order to foster and add credit to any proposed approach to address the issue and to facilitate the proposed methods endorsement by the scientific and technical community.

Author details

Luciano Burgazzi
Reactor Safety and Fuel Cycle Methods Technical Unit, ENEA, Italian National Agency for New Technologies, Energy and Sustainable Economic Development, Bologna, Italy

9. References

[1] IAEA TEC-DOC-626, 1991. Safety Related Terms for Advanced Nuclear Power Plants. September 1991.

[2] IAEA TEC DOC-1474, 2005. Natural Circulation in Water Cooled Nuclear power Plants. *Phenomena, models, and methodology for system reliability assessments,* November 2005.

[3] IAEA TECDOC-1624, 2009. Passive Safety Systems and Natural Circulation in Water Cooled Nuclear Power Plants. November 2009

[4] United States Nuclear Regulatory Commission's (U.S. NRC) Reactor Safety Study (WASH-1400, 1975).

[5] Zio, E., Pedroni, N., 2009. Building Confidence in the Reliability Assessment of Thermal hydraulic Passive Systems. Reliability Engineering and System Safety, 94, 268-281.

[6] Jafari, J., D'Auria F., et al., 2003. Reliability Evaluation of a Natural Circulation System. Nuclear Engineering and Design 224, 79–104.

[7] Marques, M., Burgazzi L., et al., 2005. Methodology for the Reliability Evaluation of a Passive System and its Integration into a Probabilistic Safety Assessment. Nuclear Engineering and Design 235, 2612-2631.

[8] Lorenzo G., et al., Assessment of an Isolation Condenser of an Integral Reactor in View of Uncertainties in Engineering Parameters, Science and technology of Nuclear Installations, Volume 2011, Article ID 827354, 9 pages

[9] Apostolakis G., Pagani L. and Hejzlar, P., 2005. The Impact of Uncertainties on the Performance of Passive Systems. Nuclear Technology 149, 129–140

[10] Apostolakis G., Mackay F., and. Hejzlar P, 2008. Incorporating Reliability Analysis into the Design of Passive Cooling System with an Application to a Gas-Cooled Reactor. Nuclear Engineering & Design 238, 217-228

[11] Burgazzi, L., 2007a. Addressing the Uncertainties related to Passive System Reliability. Progress in Nuclear Energy 49, 93-102.

[12] Burgazzi, L., 2002. Passive System Reliability Analysis: a Study on the Isolation Condenser, Nuclear Technology 139, 3-9.

[13] Burgazzi, L., 2003. Reliability Evaluation of Passive Systems through Functional Reliability Assessment, Nuclear Technology 144, 145-151.

[14] Burgazzi, L. 2007b. Thermal-hydraulic Passive System reliability-based design approach, Reliability Engineering and System Safety 92 (9), 1250-1257.

[15] Nayak, A.K., et al., 2008. Passive System Reliability Analysis using the APSRA Methodology. Nuclear Engineering and Design 238, 1430-1440.

[16] Burgazzi, L., 2004. Evaluation of Uncertainties related to Passive Systems Performance. Nuclear Engineering and Design 230, 93-106.

[17] Ricotti M.E., Zio E., D'Auria F., Caruso G., 2002. Reliability Methods for Passive Systems (RMPS) Study – Strategy and Results, in proceedings of the NEA CSNI/WGRISK Workshop on Passive System Reliability. A Challenge to Reliability Engineering and Licensing of Advanced Nuclear Power Plants, 146-163

[18] Burgazzi, L., 2008a. Reliability Prediction of Passive Systems based on Bivariate Probability Distributions, Nuclear Technology 161, 1-7.

[19] Burgazzi, L., 2009. Evaluation of the Dependencies related to Passive System Failure. Nuclear Engineering and Design 239, 3048-3053

[20] Burgazzi, L., 2011. Reliability Prediction of Passive Systems with Multiple Degradation Measures, Nuclear Technology 173, 153-161.

[21] USNRC, 2007. Feasibility study for a risk-informed and performance-based regulatory structure for future plant licensing. US Nuclear Regulatory Commission, NUREG-1860.

[22] IAEA, 2007. Proposal for a technology-neutral safety approach for new designs. International Atomic Energy Agency, TECDOC-1570, Vienna.

[23] Burgazzi, L., 2008b. Incorporation of Passive Systems within a PRA Framework. Proceedings of PSAM9, 9th International Probabilistic, Safety Assessment and Management Conference, Hong Kong, 18-23 May 2008.

[24] Mercurio, D., Podofillini, L., Zio, E., Identification and Classification of Dynamic Event Tree Scenarios via Possibilistic Clustering: Application to a Steam Generator Tube Rupture Event. Accident Analysis and Prevention 41 (2009), 1180–1191

[25] Matzie, R. A. and Worrally, A., The AP1000 reactor—the Nuclear Renaissance Option. Nuclear Energy, 2004, 43, No. 1, Feb., 33–45

[26] Burgazzi, L., 2008c. About Time-variant Reliability Analysis with Reference to Passive Systems Assessment. Reliability Engineering and System Safety 93, 1682-1688.

[27] JiYong Oh and Golay, M., 2008. Methods for Comparative Assessment of Active and Passive Safety Systems with respect to Reliability, Uncertainty, Economy and Flexibility. Proceedings of PSAM9, 9th International Probabilistic, Safety Assessment and Management Conference Hong Kong, 18-23 May 2008.

Chapter 4

Geological Disposal of Nuclear Waste: Fate and Transport of Radioactive Materials

Prabhakar Sharma

Additional information is available at the end of the chapter

1. Introduction

Nuclear power plants use nuclear fission for generating tremendous amount of heat for the production of electrical energy. Currently, there are many nuclear power plants in operation worldwide, which produces high-level nuclear wastes at the same time. Nuclear wastes are being produced as by-product of nuclear processes, like nuclear fission (spent fuel) in nuclear power plants, the radioactive elements left over from nuclear research projects and nuclear bomb production. The management and disposal of these previously stored and continuously generated nuclear wastes is a key issue worldwide. A huge amount of radioactive wastes have been stored in liquid and solid form from nuclear electricity/bomb production plants from several decades at different locations in the world. For example, the Hanford Site is a most decommissioned nuclear production complex on the Columbia River in the U.S. state of Washington, operated by the United States federal government as shown in fig 1 [21, 28, 50]. Hanford was the first large-scale plutonium production reactor in the world. The Hanford site represents approximately two-thirds of the nation's high-level radioactive waste by volume [28].

Radioactive/nuclear wastes are specific or mixture of wastes which contain radioactive chemical elements that can not be used for further power production and need to be stored permanently/long term in environmentally safe manner [63]. The ultimate disposal of these vitrified radioactive wastes or spent fuel elements requires their complete isolation from the environment. One of the most favorite method is disposal in dry and stable geological formations approximately 500 meters deep. Recently, several countries in Europe, America and Asia are investigating sites that would be technically and publicly acceptable for deep geological storage of nuclear wastes. For example, a well designed geological storage of nuclear waste from hospital and research station is in operation at relatively shallow level in Sweden and a permanent nuclear repository site is planning to be built at deep subsurface system for nuclear spent fuel in Sweden in order to accommodate the stored and running nuclear waste from ten operating nuclear reactors which produce about 40 percent of Sweden's electricity (In Sweden, the responsibility for nuclear waste management has been

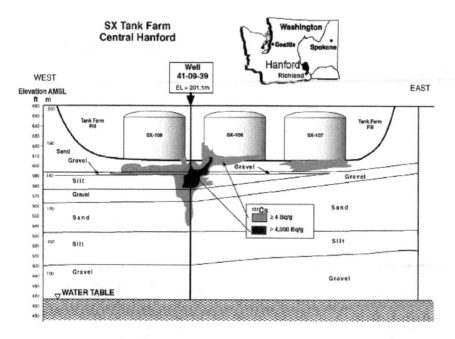

Figure 1. An example of underground storage of radioactive waste and leakage into subsurface system at Hanford site, Richland, WA, USA (from McKinley et al. 2001) [50].

transferred in 1977 from the government to the nuclear industry, requiring reactor operators to present an acceptable plan for waste management with a so called absolute safety to obtain an operating license. The conceptual design of a permanent repository was determined by 1983, calling for a placement of copper-clad iron canisters in a granite bedrock about 500 m underground, below the water table known as the KBS-3 method, an abbreviation of kärnbränslesäkerhet, nuclear fuel safety. Space around the canisters will be filled with bentonite clay. On June 3rd 2009, Swedish government choose a location for deep level waste site at Östhammar, near Forsmark nuclear power plant.).

The recent accident in 2011 in nuclear power plant in Fukushima, Japan due to Tsunami has caused release of underground stored radioactive elements/wastes into the subsurface system. This is a big concern for clean-up operation as they can migrate to farther locations with pore water flow of subsurface system and can create big environmental disaster. It has led to re-thinking of researcher and responsible organizations for protecting their underground stored radioactive wastes and implementing multi-protection mechanisms for deep geological storage of the hazardous radioactive wastes. In the event of accidental release/leakage of radioactive materials into the subsurface system, there is a possibility of its migration with the soil-pore water flow and to be transported to the surface and groundwater bodies as shown in fig 1 [21, 50, 76]. Furthermore, some radioactive contaminants do not move through soil pores in dissolved form but rather attach strongly to fine soil particles (1 nm to 1 μm size, commonly called "colloid"). These contaminant-attached soil

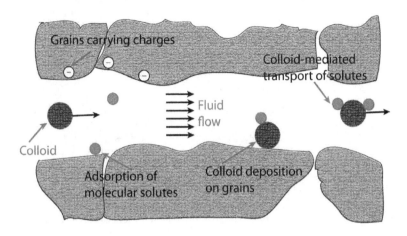

Figure 2. Flow and transport of colloidal particles attached with possible radioactive contaminants.

particles may themselves become mobile and move through the soil and eventually reach the water bodies (fig 2). This is commonly known as colloid-facilitated contaminant transport [9, 11, 14, 16, 20, 23, 41, 52, 56]. Colloids are a ubiquitous component of subsurface systems and play an important role in radioactive contaminant fate and transport. Mobile colloidal particles in the subsurface can enhance the movement of otherwise immobile radioactive contaminants attached with colloidal particles or colloids can be the radioactive elements themselves [11, 20, 23, 33, 34]. The colloidal particles can be transported in the groundwater through soil solution as affected by physico-chemical condition of the surrounding medium and colloids (fig 2). The understanding of colloid transport is of significant interest for the protection of subsurface environment from contamination by intentional or unintentional release of nuclear wastes.

In the infiltration/rainfall events, the colloidal size radioactive particles or radioactive elements attached with mobile colloids would be transported to the groundwater through unsaturated porous media, where gaseous phase can play a critical role in association with the liquid and solid phases [13, 68, 82]. Several mechanisms are responsible for colloid transport in unsaturated zone in addition to that of saturated zone, such as, liquid-gas interface capture, solid-liquid-gas interface capture, liquid-film straining, and storage in immobile liquid zones [13, 18, 27, 44, 51, 68, 70, 79, 85, 87]. The strong force (capillary force) associated with the moving liquid-gas interfaces led to particle mobilization in the natural subsurface environment. As the water content decreases, a thin film of liquid forms over the grain surfaces and in the pendular rings (smaller pores). Phenomenon of colloid deposition on these liquid film and pendular ring created between the pore spaces had different opinion in different literature [73, 81]. This chapter will review all the possible mechanisms responsible for attachment of colloids in the partially saturated system. The discrepancies in literature about colloid removal and deposition mechanisms at different locations in three phase system will also be discussed to guide the researcher and decision making bodies for designing deep geological storage for storing nuclear wastes (to ensure uninterrupted and cheap nuclear power generation) and to combat the extreme situation of their release into subsurface systems through unsaturated zone and protecting the natural water bodies and environment from radioactive contamination.

2. Mechanism of colloid attachment

The colloid retention in saturated porous media is primarily controlled by attachment at the solid-liquid interface in relation to the surface properties of the solid and background solution, which has been well documented in literature [37, 42, 46, 48, 49, 60, 65, 69]. Whereas the presence of gaseous phase in the unsaturated subsurface system introduces an additional mechanism for colloid retention. Although several steps has been taken to enhance the understanding of mechanisms responsible for colloid transport and retention through unsaturated porous media, there is a need to put extra effort in this area for better understanding [4, 24, 42, 47, 55]. In the unsaturated porous media, the additional mechanisms (compared to saturated system) for colloid transport were reported as: colloid captured at the liquid-gas interface [1, 12, 43, 44, 54, 66–68, 70, 72, 80, 83], colloid captured due to straining [4, 7, 74, 78], the colloid captured at solid-liquid-gas interface [10, 17, 18, 27, 51, 87, 88], and colloid storage in immobile zone [15, 25, 26, 61]. The flow chart lists the above four retention mechanisms (fig 3). The colloids trapped due to different mechanisms, as mentioned in the flow chart, govern the movement of colloidal/nano-size particles in a porous media (Fig. 4). The figure 4 shows the example of the colloid captured by liquid-gas interface, solid-liquid-gas triple point, straining, immobile zone, and solid-liquid interface. Many of the colloid retention mechanisms are still poorly understood and debating [73, 81]. To improve our knowledge and understanding about the fate of radioactive particles (alone or attached with colloidal particle) in unsaturated porous media, the colloid capture mechanisms are discussed in detail below.

2.1. Attachment at the liquid-gas interfaces

It has been stated in the past that the moving liquid-gas interface plays an important role in colloid mobilization in unsaturated porous medium [1, 12, 43, 44, 54, 66–68, 70, 72, 80, 83]. A considerable amount of colloids were captured at the liquid-gas interfaces and moved with the infiltration front depending on flow velocity and the solution ionic strength [68]. This has been verified by numerical solution of the Young-Laplace equation that expanding water film can lift the subsurface colloids from the mineral surfaces [66]. The detachment of sub-micron sized particles from initially wet solid surfaces had been investigated by air-bubble experiments to understand the strength of moving liquid-gas interfaces [30–32, 45, 53]. In a direct visualization experiments, it had been found that a significant number of colloids were detached from initially dried solid surfaces by the moving liquid-gas interface and remain attached to the liquid-gas interfaces (Fig 5) [70]. The irreversible nature of colloid attachment from the liquid-gas interface has been observed earlier, which validate the strength of moving liquid-gas interface [1, 80].

Three consequent steps might occur in the colloid detachment from a solid surface and its attachment to the liquid-gas interface. These are interception of the particle, attachment or thinning of the liquid film in between the particle and the liquid-gas interface, and

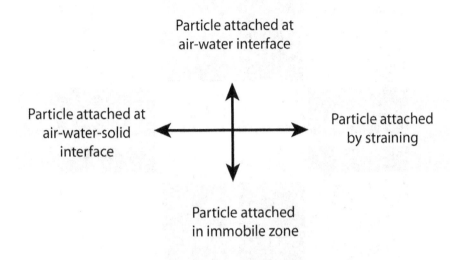

Figure 3. Different possible colloid retention mechanisms in unsaturated porous media.

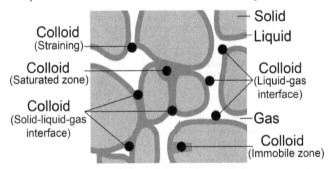

Figure 4. Attachment of colloid in three-phase system.

stabilization of the particle on the liquid-gas interface [22, 30, 67, 70]. The total detachment probability (P_{det}) is defined as [22]:

$$P_{det} = P_{int} \times P_{att} \times P_{sta} \tag{1}$$

where P_{int} is the interception/colloision probability, P_{att} is the attachment probability, and P_{sta} is the stability probability. P_{int} depends on the actual number of colloids intercepted by the moving liquid-gas interface. It depends on the velocity and direction of the liquid-gas interface with respect to the solid surface. P_{att} depends on velocity of the liquid-gas interface. It could be zero, means no colloid detachment from solid surface, if interface contact time will be less than induction time. The induction time is time to form a three-phase contact line by thinning the liquid film between the particle and the liquid-gas interface. P_{sta} can be assumed

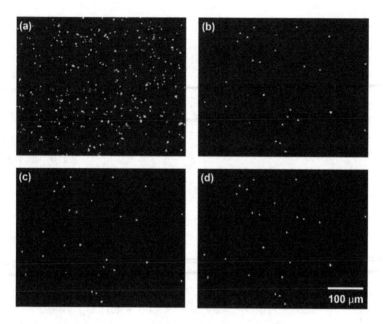

Figure 5. Detachment of amine-modified microspheres from glass slide after moving the liquid-gas interface: (a) no interface movement, (b) 1 interface movements, (c) 2 interface movements, and (d) 3 interface movements.

as one because of irreversible nature of colloids attached on the liquid-gas interface, i.e., particles remain attached to the interface. Thus, the above equation (1) shows the importance of velocity (specially in the range of porous media velocity) on detachment of colloid by the moving liquid-gas interface. Sharma et al [70] tested the velocity effect in the range of 0.4 to 400 cm h^{-1} on colloid detachment from the solid surface. They found that colloid detachment from the solid surface was more at lower interface velocity. A similar observation has been made for particle detachment by air-bubble moving at higher speed (> 8 cm h^{-1}) [30, 31].

For the transport and mobilization of radioactive materials in colloidal size or its attachment with colloidal particle, the balance among electrostatic, hydrodynamic, and capillary forces are responsible for attraction of particle towards the liquid-gas interface [29, 67, 68, 70, 72, 79]. If capillary force dominates then the colloidal particles attracted towards the liquid-gas interface and if electrostatic dominates then the colloidal particles remain stay over the grain surface. The hydrodynamic forces may be neglected for the colloidal size particles [57, 58, 64]. Figure 6 shows the force balance between electrostatic force and the capillary force for hydrophilic and hydrophobic particle attached with the solid surface when liquid-gas interface moved in the upward direction. The attachment force (F_{att}) is the sum of electrostatic force and van der Waals force acting toward the solid surface whereas the detachment force (F_{det}) is the horizontal component of the capillary force, which is responsible for detachment of colloids from the solid surface and attachment into the liquid-gas interface as shown below in detail. Colloids detach from the solid surface only if $F_{det} > F_{att}$. The detail of force balance calculation are shown below:

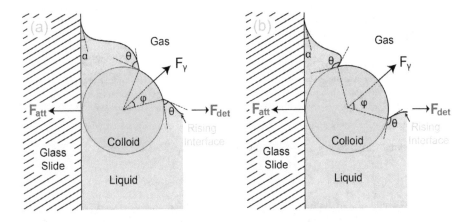

Figure 6. Schematic of forces exerted on an adhered particle: (a) hydrophilic and (b) hydrophobic particle, in contact with a liquid-gas interface. Gravity and buoyancy forces are neglected. (F_{det}: detachment force, F_{att}: attachment force, F_{fl}: surface tension force, θ: contact angle for colloids, α: contact angle for glass surface, ϕ: filling angle).

2.1.1. DLVO forces

The DLVO profiles for the colloids and their interaction with the glass surface were calculated according to [35]:

$$\Delta G_{el} = 64\pi\epsilon R \left(\frac{kT}{ze}\right)^2 \left[\tanh\left(\frac{ze\psi_{0,1}}{4kT}\right)\right]\left[\tanh\left(\frac{ze\psi_{0,2}}{4kT}\right)\right]\exp(-\kappa h) \qquad (2)$$

where ΔG_{el} is the electrostatic interaction energy, ϵ is the dielectric permittivity of the medium, R is the radius of the colloids, k is the Boltzmann constant, T is the absolute temperature; z is the ion valence, e is the electron charge, $\psi_{0,1}$ and $\psi_{0,2}$ are surface potential of the colloids and the glass slide respectively, which are taken as the colloid and the glass ζ-potentials, h is the separation distance, κ is the inverse Debye-Hückel length, $\kappa = \sqrt{\frac{e^2 \sum n_j z_j^2}{\epsilon kT}}$, where n_j is the number concentration of the ions in solution, and z_j is the ion valence.

The van der Waals interaction energy was calculated by [36]:

$$\Delta G_{vdw} = -\frac{AR}{6h}\left[1 - \frac{5.32h}{\lambda_0}\ln\left(1 + \frac{\lambda_0}{5.32h}\right)\right] \qquad (3)$$

where A is the effective Hamaker constant of colloid-water-glass system, and λ_0 is a characteristic length of 100 nm. The effective Hamaker constant ($A = A_{123}$) was calculated using individual Hamaker constant of colloid, water, and glass [38].

$$A_{123} = (\sqrt{A_{11}} - \sqrt{A_{22}})(\sqrt{A_{33}} - \sqrt{A_{22}}) \qquad (4)$$

where A_{11} is the Hamaker constant of the colloids, A_{22} is the Hamaker constant of the fluid, and A_{33} is the Hamaker constant of the glass.

Polystyrene colloids	Diameter[a] (μm)	Contact angle[b] (deg)	Surface charge[a] (meq/g)	CaCl₂ conc. (mM)	pH	Experimental Conditions			
						Electrophoretic mobility[c] (–) (μm/s)/(V/cm)	ζ-potential[d] (mV)	Colloid conc. (particles/L)	
Amino-modified	1.0±0.02	20.3±1.9	0.1047	6	5.9	0.15±0.02	1.9±0.2	7.2×10⁸	

[a]Values provided by manufacturer. [b]Measured with a goniometer (DSA 100, Krüss, Hamburg, Germany). [c]Measured with a ZetaSizer 3000HSa (Malvern Instruments Ltd., Malvern, UK) at the electrolyte concentration and pH indicated in the table. [d]Obtained from measured electrophoretic mobilities using the von Smoluchowski equation [38].

Table 1. Selected properties of polystyrene colloids and suspension chemistry used in the experiments.

Finally, the total DLVO forces were calculated as:

$$F_{DLVO} = \frac{d}{dh}\left(\Delta G_{tot}\right) = \frac{d}{dh}\left(\Delta G_{el} + \Delta G_{vdw}\right) \tag{5}$$

To see an example of particle detachment from initially dried glass surfaces, [70] performed experiments by selecting different types of colloids with their modified surface properties. Parameters for the DLVO calculations for one of the colloids are shown in Table 1, and the Hamaker constant was chosen as that for a polystyrene-water-glass system (polystyrene: $A_{11} = 6.6 \times 10^{-20}$ J, water: $A_{22} = 3.7 \times 10^{-20}$ J, glass: $A_{33} = 6.34 \times 10^{-20}$ J; all data taken from Israelachvili [39]; the combined Hamaker constant calculated with equation 4 is $A_{123} = 3.84 \times 10^{-21}$ J).

2.1.2. Surface tension forces

The total force exerted by a moving liquid-gas interface on a colloidal particle is the sum of gravity, buoyancy, and interfacial forces. However, the gravity and buoyancy forces can be neglected for small particles with radii < 500 μm [57, 58, 64, 70]. In experimental setup, when the liquid-gas interface moves in upward direction over the vertically mounted glass slide, the horizontal component of surface tension force (F_γ) is the detachment force (F_{det}) which is opposed by the DLVO force (F_{att}) (Figure 6). The detachment force (the maximum horizontal surface tension force) can be calculated by [45, 57, 58, 64]:

$$F_{det} = 2\pi R\gamma \sin^2\left(\frac{\theta}{2}\right)\cos\alpha \tag{6}$$

where R is the radius of the particle, γ is the surface tension of liquid, and θ and α are the advancing contact angles for colloids and the glass slide, respectively.

The experiments were conducted using hydrophilic and hydrophobic modified surface and positively and negatively charged colloids attached over the negatively charged glass slide to estimate the number of colloids removed by moving liquid-gas interface [70]. Colloids over the glass slide were visualized using laser scanning confocal microscopy. Figure 5 shows an example of confocal images before and after moving the liquid-gas interfaces over the glass slide. The figure shows that a considerable amount of colloids were removed by the passage of the first liquid-gas interface (Fig 5a,b), however more number of passages of liquid-gas interface did not affect the colloid left after the first interface movement (Fig 5c,d). This was caused because some of the particles might have attached in the primary energy minimum from the glass slide, so F_{att} would be much larger than F_{det} for those particles [70]. From the DLVO calculations using eq 5, there is a favorable attachment for amino-modified

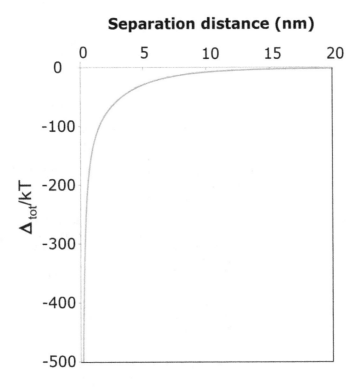

Figure 7. DLVO profile of amino-modified colloid (at given condition in Table 1).

microspheres i.e., a strong attractive force between colloids and the glass surface (fig 7). Shang et al [67] had recently studied the total force balance exerted on a particle passing through a liquid-gas interface. They considered different shape and size (1 to 6 μm) of particles of different surface properties to pass through a liquid-gas interface and measured the forces exerted on the particle over time using a tensiometer and compared with their theocratical force balance calculations. It has been observed that the liquid-gas interface due to capillary force generates a strong repulsion of particles from the stationary surfaces when water film expands or move through the solid surfaces [67]. In order to detach a particle from the solid surfaces, a liquid film larger than the particle diameter must build up around the particle so that the repulsive capillary force can dominate and a lift can occur.

The above discussion based on force balance complimented with visualization experiments imply that moving liquid-gas interface tends to dominate colloid movement during water infiltration into soils and sediments. The strong force associated at the liquid-gas interface can overcome colloid aggregation and settling, which otherwise dominate colloid dispersion and mobility in porous media. The strong affinity of colloidal particle towards the liquid-gas interface may also be applied in remediation technology, as the inert gases in the form of gas-bubbles can be injected in soils or aquifers to preferentially mobilize colloidal size radioactive contaminants.

2.2. Attachment of colloid by straining

The infiltration and drainage scenarios are quite common in the event of rainfall and drying on the unsaturated zone of the subsurface system. This processes can complex the mobilization of radioactive particles in the upper layers if there is any spill or leakage of those material. In the unsaturated zone as the water drain, sorb or evaporate, the water thickness over the solid surface becomes thinner and thinner; and once the water film becomes thinner than colloid diameters that mechanism is called water film staining. In this case, a strong force exerted on the colloid towards the solid surface which is called capillary force [70, 77, 78, 90]. Other possibility of film straining was explained by the colloid trapped in the pendular rings (smaller pores) region separated by thin water films from the remaining fluids [4, 78], which can be remobilized after expanding the water films [26, 61]. The straining of colloids also happen if the pore sizes are smaller than colloid size. This phenomenon commonly occurs in the saturated zone which can also happen in the unsaturated zone.

Different types of straining mechanism for colloid attachment were studied by Bradford group and others [2–8, 19, 40, 59, 71, 74, 84–86]. Figure 8 shows the different types of straining locations for colloids in the saturated and unsaturated media. Colloids trapped at the intersection point of two solid grains in the saturated systems at location 1 by single and 2 by multiple colloidal particles are also called wedging [40] and bridging [59] respectively. The straining of single particle (location 1) occurs if the pore spaces in a porous medium are smaller than the colloid diameter, which is a common phenomenon applied in mechanical filtration [49]. However, straining of multiple particles (location 2) occur as a result of aggregation of colloidal particles in the solution, although the pore space is larger than the single colloid diameter.

In addition, straining of colloids in the saturated system also depends on solution properties, colloid size, colloid shape, colloid size distribution as well as grain size and heterogeneity [2, 3, 5, 6, 71, 85, 86]. The straining of colloids were more dominant for large, irregular shape, and multi-disperse colloids [84–86]. Straining of colloids in the unsaturated porous media become very complex due to the presence of gaseous phase. The capillary force controls the distribution of liquid and gas phases in the pores. As the amount of liquid decreased from the porous medium, the liquid form a film over the solid surface or retain the smaller pores due to strong capillary forces and the larger pores are filled with gases [75]. Straining behavior of colloids due to pore sizes in the unsaturated systems were not studied yet, however few efforts had been taken on straining of colloid by liquid film [62, 78] and the colloid attachment at the solid-liquid-gas triple point [17, 18, 27, 51, 87, 88]. The example of colloids retained at the solid-liquid-gas triple point are shown in Fig 8 at location 3, which has been discussed in detail in the next section. Straining of colloids in the unsaturated porous media due to liquid film occurred if the liquid thickness is smaller than colloid diameters (location 4 in Fig 8). [78] concluded, using different size of colloids and by changing flow velocity, that colloids with smaller diameter than water film thickness passed easily but colloids bigger than film thickness were trapped on the water film.

2.3. Attachment at the solid-liquid-gas interfaces

The contact point of solid-liquid and liquid-gas is called solid-liquid-gas interface. Steenhuis and coworkers used infiltration chambers, light source, and imaging system (camera setup or confocal microscope) to study the colloids attached at so called air/water-meniscus/solid

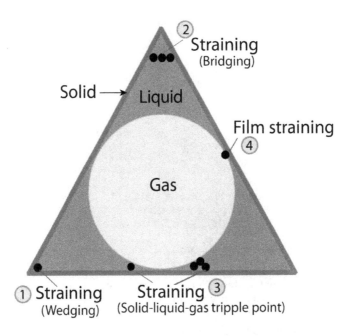

Figure 8. Attachment of colloids in porous media due to straining.

(AW$_m$S) interface (inside the narrow portion of the pendular ring) in the unsaturated porous media [17, 18, 27, 51, 87, 88]. They established from their visualization experiments that colloids tend to accumulate at AW$_m$S interface in the unsaturated porous media. They reiterated that the hydrophilic colloids were deposited at AW$_m$S interface whereas hydrophobic colloids deposited at the solid-liquid interface, but none of them were present at the liquid-gas interface [17, 18]. The capillary force calculation had given the theoretical explanation why the colloids attracted towards AW$_m$S interface [27]. Their force calculation showed that the colloid retention at AW$_m$S interface is only possible for hydrophilic colloids with contact angle less than 45^0 for sand grain medium (Fig 9). So the colloid were not attached at the AW$_m$S interfaces in the friction coefficient were bellow the tangent of contact angle.

Contrarily, the deposition of colloids were found at the liquid film (liquid-gas interface) from glass micromodel experiments and modeling studies [78, 80]. In another visualization studies, colloids accumulation were found at thin films outside the pendular ring, which was air-water interface not connected with the solid grains [26]. A column and micromodel experiments and thermodynamic calculations showed that colloids were most likely to be retained near the sediments of liquid-gas interface i.e., solid-liquid-gas interface attachment [10]. These discrepancies in the literature between the colloid attachment mechanisms due to the presence of the solid-liquid-gas interface in the unsaturated porous media had been debated [73, 81]. [81] argued that the possible cause of colloid attachment at AW$_m$S interface was evaporation

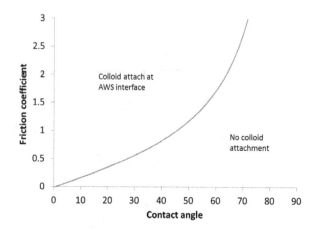

Figure 9. Relationship between grain contact angle and friction coefficient for colloid retention at the AW$_m$S interface.

in the chamber, drying of thin water film over the grain surface, and advection of colloids to AW$_m$S contact line. But it had been refuted by Steenhuis et al [73].

2.4. Attachment at the immobile zones

In the partially saturated systems, colloids were found to be captured into stagnant/immobile zone. There was evidence of exchange of colloids between immobile and mobile zone due to long breakthrough curve tailing on colloid transport through unsaturated porous media [15, 25, 61, 68]. In a visualization study, it was found that colloids present in the immobile zone at the liquid-gas interface were not moved to mobile zone in steady flow, but the exchange of colloids between immobile and mobile zone occurred in varied flow rate [26]. The exchange of colloids between mobile and immobile zones were likely controlled by slow advection in addition to diffusion. The occurrence of larger quantity of colloids from unsaturated column studies were found in transient flow condition due to movement of colloids present in immobile zone [61, 66, 89]. All these studies indicated that the colloid can be attached in the immobile zone created by heterogeneity of the medium and by the presence of gaseous phase, which could be remobilized in the large rainfall and infiltration events.

3. Conclusions and future directions

The study of colloid fate and transport in important as there is strong affinity of radioactive contaminants to attach with the moving colloidal particles or radioactive elements can fall under colloidal size range. In subsurface systems (like soils and sediments) moving air-water interfaces are common, e.g., during infiltration and drainage of water, air and water displace each other in continuous cycles. Such moving air-water interfaces have a profound effect on detachment of colloids from surfaces. Several research efforts had been made to understand the mechanism of colloid retention and mobilization in unsaturated porous media. The possible cause for colloid attachment in the presence of gaseous phase are discussed in this chapter. As discussed in this chapter, it is difficult to draw firm conclusions about the colloid capture locations in unsaturated porous media. The column experiments, modeling techniques, and visualization studies reveal a number of possible mechanisms of colloid retention and deposition in the partially saturated systems. It is likely that the colloidal particles attached with the solid grain can be removed by moving liquid-gas interface and then colloids can be either deposited and restrained from further moving due to different types of straining, solid-liquid-gas interface capture, and the presence of immobile zone of heterogeneous medium or remain attached at the liquid-gas interfaces.

The strong attachment of radioactive particles to liquid-gas interfaces leading to removal of stationary surfaces offers opportunities for management of subsurface systems in terms of flow and transport. Infiltration fronts in soils can be readily generated by flooding, for instance, and radioactive particle can be effectively "washed" out of a soil profile. Air-bubbles in the form of N_2 or other inert gases may be injected in soils or aquifers to preferentially mobilize and remove radioactive contaminants. Such techniques offer ways to enhance the mobility of otherwise immobile particles in the vadose zone and in groundwater. The results from this study point to the relevance of moving air-water interfaces for nuclear waste mobilization and transport in the vadose zone. Such moving air-water interfaces are common in soils and near-surface sediments, where rainfall, snow melt, or irrigation cause infiltration and drainage. Current theory for colloid transport in unsaturated porous media does not consider the effect of moving air-water interfaces for release of contaminants. Evidently, the colloid removal, transport, and deposition mechanisms remain a fertile area of research with much still left to investigate and opportunities for progress in both theory and experiments that are likely to have significant practical impact in vadose zone fate and transport of colloid attached contaminants for better understanding of any radioactive contamination transport from the release point to farther location.

4. Abbreviations

AW_mS: Air/water-meniscus/solid
DLVO: Derjaguin, Landau, Verwey and Overbeek
KBS: Kärnbränslesäkerhet

Author details

Prabhakar Sharma
Department of Earth Sciences, Uppsala University, Uppsala, Sweden

5. References

[1] Abdel-Fattah, A. I. & El-Genk, M. S. [1998]. On colloidal particle sorption onto a stagnant air-water interface, *Adv. Colloid. Interface Sci.* 78: 237–266.
[2] Bradford, S. A. & Bettahar, M. [2005]. Straining, attachment, and detachment of *Cryptosporidium* oocysts in saturated porous media, *J. Environ. Qual.* 34: 469–478.
[3] Bradford, S. A., Bettahar, M., Šimůnek, J. & van Genuchten, M. T. [2004]. Straining and attachment of colloids in physically heterogeneous porous media, *Vadose Zone J.* 3: 384–394.

[4] Bradford, S. A. & Torkzaban, S. [2008]. Colloid transport and retention in unsaturated porous media: A review of interface-, collector-, and pore-scale processes and models, *Vadose Zone J.* 7: 667–681.

[5] Bradford, S. A., Šimůnek, J., Bettahar, M., Tadassa, Y. F., van Genuchten, M. T. & Yates, S. R. [2005]. Straining of colloids at textural interfaces, *Water Resour. Res.* 11: W10404, doi:10.1029/2004WR003675.

[6] Bradford, S. A., Šimůnek, J., Bettahar, M., van Genuchten, M. T. & Yates, S. R. [2006]. Significance of straining in colloid deposition: Evidence and implications, *Water Resour. Res.* 42: W12S15, doi:10.1029/2005WR004791.

[7] Bradford, S. A., Šimůnek, J. & Walker, S. L. [2006]. Transport and straining of *E. coli* O157:H7 in saturated porous media, *Water Resour. Res.* 42: W12S12, doi:10.1029/2005WR004805.

[8] Bradford, S. A., Yates, S. R., Bettahar, M. & Šimůnek, J. [2002]. Physical factors affecting the transport and fate of colloids in saturated porous media, *Water Resour. Res.* 38: 1327, doi:10.1029/2002WR001340.

[9] Chawla, F., Steinmann, P., Loizeau, J. L., Hossouna, M. & Froidevaux, P. [2010]. Binding of 239pu and 90sr to organic colloids in soil solutions: Evidence from a field experiment, *Environ. Sci. Technol.* 44: 8509–8514.

[10] Chen, G. & Flury, M. [2005]. Retention of mineral colloids in unsaturated porous media as related to their surface properties, *Colloids Surf. Physicochem. Eng. Aspects* 256: 207–216.

[11] Chen, G., Flury, M. & Harsh, J. B. [2005]. Colloid-facilitated transport of cesium in variable-saturated Hanford sediments, *Environ. Sci. Technol.* 39: 3435–3442.

[12] Chen, L., Sabatini, D. A. & Kibbey, T. C. G. [2008]. Role of the air-water interface in the retention of TiO_2 nanoparticles in porous media during primary drainage, *Environ. Sci. Technol.* 42: 1916–1921.

[13] Cheng, T. & Saiers, J. E. [2009]. Mobilization and transport of in situ colloids during drainage and imbibition of partially saturated sediments, *Water Resour. Res.* 45: W08414, doi:10.1029/2008WR007494.

[14] Cheng, T. & Saiers, J. E. [2010]. Colloid-facilitated transport of cesium in vadose-zone sediments: The importance of flow transients, *Environ. Sci. Technol.* 44: 7443–7449.

[15] Cherrey, K. D., Flury, M. & Harsh, J. B. [2003]. Nitrate and colloid transport through coarse hanford sediments under steady state, variably saturated flow, *Water Resour. Res.* 39: 1165, doi:10.1029/2002WR001944.

[16] Crancon, P., Pili, E. & Charlet, L. [2010]. Uranium facilitated transport by water-dispersible colloids in field and soil columns, *Sci. Total Environ.* 408: 2118–2128.

[17] Crist, J. T., McCarthy, J. F., Zevi, Y., Baveye, P., Throop, J. A. & Steenhuis, T. S. [2004]. Pore-scale visualization of colloid transport and retention in party saturated porous media, *Vadose Zone J.* 3: 444–450.

[18] Crist, J. T., Zevi, Y., McCarthy, J. F., Troop, J. A. & Steenhuis, T. S. [2005]. Transport and retention mechanisms of colloids in partially saturated porous media, *Vadose Zone J.* 4: 184–195.

[19] Cushing, R. S. & Lawler, D. F. [1998]. Depth filtration: Fundamental investigation through three-dimensional trajectory analysis, *Environ. Sci. Technol.* 32: 3793–3801.

[20] Czigany, S., Flury, M., Harsh, J. B., Williams, B. C. & Shira, J. M. [2005]. Suitability of fiberglass wicks to sample colloids from vadose zone pore water, *Vadose Zone J.* 4: 175–183.

[21] Dai, M., Buesseler, K. O. & Pike, S. M. [2005]. Plutonium in groundwater at the 100k-area of the U.S. DOE Hanford site, *J. Contam. Hydrol.* 76: 167–189.

[22] Dai, Z., Fornasiero, D. & Ralston, J. [1999]. Particle-bubble attachment in mineral flotation, *J. Colloid Interface Sci.* 217: 70–76.

[23] Flury, M., Mathison, J. B. & Harsh, J. B. [2002]. *In situ* mobilization of colloids and transport of cesium in Hanford sediments, *Environ. Sci. Technol.* 36: 5335–5341.

[24] Flury, M. & Qiu, H. [2008]. Modeling colloid-facilitated contaminant transport in the vadose zone, *Vadose Zone J.* 7: 682–697.

[25] Gamerdinger, A. P. & Kaplan, D. I. [2001]. Physical and chemical determinants of colloid transport and deposition in water-unsaturated sand and Yucca Mountain tuff material, *Environ. Sci. Technol.* 35: 2497–2504.

[26] Gao, B., Saiers, J. E. & Ryan, J. N. [2006]. Pore-scale mechanisms of colloid deposition and mobilization during steady and transient flow through unsaturated granular media, *Water Resour. Res.* 42: W01410, doi:10.1029/2005WR004233.

[27] Gao, B., Steenhuis, T. S., Zevi, Y., Morales, V. L., Nieber, J. L., Richards, B. K., McCarthy, J. F. & Parlange, J. Y. [2008]. Capillary retention of colloids in unsaturated porous media, *Water Resour. Res.* 44: W04504, doi:10.1029/2006WR005332.

[28] Gephart, R. E. [2010]. A short history of waste management at the hanford site, *Physics Chem. Earth* 35: 298–306.

[29] Gillies, G., Kappl, M. & Butt, H. [2005]. Direct measurements of particle-bubble interactions, *Adv. Colloid. Interface Sci.* 114–115: 165–172.

[30] Gomez-Suarez, C., Noordmans, J., van der Mei, H. C. & Busscher, H. J. [1999a]. Detachment of colloidal particles from collector surfaces with different electrostatic charge and hydrophobicity by attachment to air bubbles in a parallel plate flow chamber, *Phys. Chem. Chem. Phys.* 1: 4423–4427.

[31] Gomez-Suarez, C., Noordmans, J., van der Mei, H. C. & Busscher, H. J. [1999b]. Removal of colloidal particles from quartz collector surfaces as simulated by the passage of liquid-air interfaces, *Langmuir* 15: 5123–5127.

[32] Gomez-Suarez, C., Noordmans, J., van der Mei, H. C. & Busscher, H. J. [2001]. Air bubble-induced detachment of polystyrene particles with different sizes from collector surfaces in a parallel plate flow chamber, *Colloids Surf.* 186: 211–219.

[33] Graham, M. C., Oliver, I. W., MacKenzie, A. B., Ellam, R. M. & Farmer, J. G. [2008]. An integrated colloid fractionation approach applied to the characterisation of porewater uraniumŰhumic interactions at a depleted uranium contaminated site, *Sci. Total Environ.* 404: 207–217.

[34] Graham, M. C., Oliver, I. W., MacKenzie, A. B., Ellam, R. M. & Farmer, J. G. [2011]. Mechanisms controlling lateral and vertical porewater migration of depleted uranium (DU) at two UK weapons testing sites, *Sci. Total Environ.* 409: 1854–1866.

[35] Gregory, J. [1975]. Interaction of unequal double layers at constant charge, *J. Colloid Interface Sci.* 51(1): 44–51.

[36] Gregory, J. [1981]. Approximate expressions for retarded van der Walls interaction, *J. Colloid Interface Sci.* 83(1): 138–145.

[37] Grolimund, D., Elimelech, M., Borkovec, M., Barmettler, K., Kretzschmar, R. & Sticher, H. [1998]. Transport of in situ mobilized colloidal particles in packed soil columns, *Environ. Sci. Technol.* 32: 3562–3569.

[38] Hiemenz, P. C. & Rajagopalan, R. [1997]. *Principles of Colloid and Surface Chemistry*, Marcel Dekker Inc., New York.

[39] Israelachvili, J. [1992]. *Intermolecular and Surface Forces*, Academic Press, London.

[40] Johnson, W. P., Li, X. & Yal, G. [2007]. Colloid retention in porous media: Mechanistic confirmation of wedging and retention in zones of flow stagnation, *Environ. Sci. Technol.* 41: 1279–1287.

[41] Kersting, A. B., Efurd, D. W., Finnegan, D. L., Rokop, D. J., Smith, D. K. & Thompson, J. L. [1999]. Migration of plutonium in groundwater at thenevada test site, *Nature* 397: 56–59.

[42] Kretzschmar, R., Borkovec, M., Grolimund, D. & Elimelech, M. [1999]. Mobile subsurface colloids and their role in contaminant transport, *Adv. Agron.* 66: 121–193.

[43] Lazouskaya, V. & Jin, Y. [2008]. Colloid retention at air-water interface in a capillary channel, *Colloids Surf. Physicochem. Eng. Aspects* 325: 141–151.

[44] Lazouskaya, V., Jin, Y. & Or, D. [2006]. Interfacial interactions and colloid retention under steady flows in a capillary channel, *J. Colloid Interface Sci.* 303: 171–184.

[45] Leenaars, A. F. M. & O'Brien, S. B. G. [1989]. Particle removal from silicon substrates using surface tension forces, *Philips J. Res.* 44: 183–209.

[46] Lenhart, J. J. & Saiers, J. E. [2003]. Colloid mobilization in water-saturated porous media under transient chemical conditions, *Environ. Sci. Technol.* 37: 2780–2787.

[47] McCarthy, J. F. & McKay, L. D. [2004]. Colloid transport in the subsurface: Past, present, and future challenges, *Vadose Zone J.* 3: 326–337.

[48] McCarthy, J. & Zachara, J. [1989]. Subsurface transport of contaminants, *Environ. Sci. Technol.* 23: 496–502.

[49] McDowell-Boyer, L. M., Hunt, J. R. & Sitar, N. [1986]. Particle transport through porous media, *Water Resour. Res.* 22: 1901–1921.

[50] McKinley, J. P., Zeissler, C. J., Zachara, J. M., Serne, R. J., Lindstrom, R. M., Schaef, H. T. & Orr, R. D. [2001]. Distribution and retention of Cs-137 in sediments at the Hanford Site, Washington, *Environ. Sci. Technol.* 35: 3433–3441.

[51] Morales, V. L., Gao, B. & Steenhuis, T. [2009]. Grain surface-roughness effects on colloidal retention in the vadose zone, *Vadose Zone J.* 8: 11–20.

[52] Moridis, G. J., Hu, Q., Wu, Y. S. & Bodvarsson, G. S. [2003]. Preliminary 3-D site-scale studies of radioactive colloid transport in the unsaturated zone at Yucca Mountain, Nevada, *J. Contam. Hydrol.* 60: 251–286.

[53] Noordmans, J., Wit, P. J., van der Mei, H. C. & Busscher, H. J. [1997]. Detachment of polystyrene particles from collector surfaces by surface tension forces induced by air-bubble passage through a parallel plate flow chamber, *J. Adhesion Sci. Technol.* 11: 957–969.

[54] Oettel, M. & Dietrich, S. [2008]. Colloidal interactions at fluid interfaces, *Langmuir* 24: 1425–1441.

[55] Ouyang, Y., Shinde, D., Mansell, R. S. & Harris, W. [1996]. Colloid enhanced transport of chemicals in subsurface environments: A review, *Crit. Rev. Environ. Sci. Technol.* 26: 189–204.

[56] Pedrot, M., Dia, A., Davranche, M., Coz, M. B., Henin, O. & Gruau, G. [2008]. Insights into colloid-mediated trace element release at the soil/water interface, *J. Colloid Interface Sci.* 325: 187–197.

[57] Pitois, O. & Chateau, X. [2002]. Small particles at a fluid interface: effect of contact angle hysteresis on force and work of detachment, *Langmuir* 18: 9751–9756.

[58] Preuss, M. & Butt, H. [1998]. Measuring the contact angle of individual colloidal particles, *J. Colloid Interface Sci.* 208: 468–477.

[59] Ramachandran, V. & Fogler, H. S. [1999]. Plugging by hydrodynamic bridging during flow of stable colloidal particles within cylindrical pores, *J. Fluid Mech.* 385: 129–156.

[60] Ryan, J. N. & Elimelech, M. [1996]. Colloid mobilization and transport in groundwater, *Colloids Surf. Physicochem. Eng. Aspects* 107: 1–56.

[61] Saiers, J. E. & Lenhart, J. J. [2003a]. Colloid mobilization and transport within unsaturated porous media under transient-flow conditions, *Water Resour. Res.* 39: 1019, doi:10.1029/2002WR001370.

[62] Saiers, J. E. & Lenhart, J. J. [2003b]. Ionic-strength effects on colloid transport and interfacial reactions in partially saturated porous media, *Water Resour. Res.* 39: 1256, doi:10.1029/2002WR001887.

[63] Salbu, B., Krekling, T. & Oughton, D. H. [1998]. Characterisation of radioactive particles in the environment, *Analyst* 123: 843–849.

[64] Scheludko, A., Toshev, B. V. & Bojadjiev, D. T. [1976]. Attachment of particles to a liquid surface (Capillary theory of flotation), *J. Chem. Soc. Faraday Trans.* I 72: 2815–2828.

[65] Sen, T. K. & Khilar, K. C. [2006]. Review on subsurface colloids and colloid-associated contaminant transport in saturated porous media, *Adv. Colloid. Interface Sci.* 119: 71–96.

[66] Shang, J., Flury, M., Chen, G. & Zhuang, J. [2008]. Impact of flow rate, water content, and capillary forces on in situ colloid mobilization during infiltration in unsaturated sediments, *Water Resour. Res.* 44: W06411, doi:10.1029/2007WR006516.

[67] Shang, J., Flury, M. & Deng, Y. [2009]. Force measurements between particles and the air-water interface: Implications for particle mobilization in unsaturated porous media, *Water Resour. Res.* 45: W06420, doi:10.1029/2008WR007384.

[68] Sharma, P., Abdou, H. & Flury, M. [2008]. Effect of the lower boundary condition and flotation on colloid mobilization in unsaturated sandy sediments, *Vadose Zone J.* 7(3): 930–940.

[69] Sharma, P., Flury, M. & Mattson, E. [2008]. Studying colloid transport in porous media using a geocentrifuge, *Water Resour. Res.* 44: W07407, doi:10.1029/2007WR006456.

[70] Sharma, P., Flury, M. & Zhou, J. [2008]. Detachment of colloids from a solid surface by a moving air-water interface, *J. Colloid Interface Sci.* 326: 143–150.

[71] Shen, C., Huang, Y., Li, B. & Jin, Y. [2008]. Effects of solution chemistry on straining of colloids in porous media under unfavorable condition, *Water Resour. Res.* 44: W05419, doi:10.1029/2007WR006580.

[72] Sirivithayapakorn, S. & Keller, A. [2003]. Transport of colloids in unsaturated porous media: a pore-scale observation of processes during the dissolution of air-water interface, *Water Resour. Res.* 39: 1346, doi:10.1029/2003WR002487.

[73] Steenhuis, T. S., McCarthy, J. F., Crist, J. T., Zevi, Y., Baveye, P. C., Throop, J. A., Fehrman, R. L., Dathe, A. & Richards, B. K. [2005]. Reply to "comments on 'pore-scale visualization of colloid transport and retention in partly saturated porous media'", *Vadose Zone J.* 4: 957–958.

[74] Torkzaban, S., Bradford, S. A., van Genuchten, M. T. & Walker, S. L. [2008]. Colloid transport in unsaturated porous media: The role of water content and ionic strength on particle straining, *J. Contam. Hydrol.* 96: 113–127.

[75] Tuller, M. & Or, D. [2001]. Hydraulic conductivity of variably saturated porous media: Film and corner flow in angular pores, *Water Resour. Res.* 37: 1257–1276.

[76] Utsunomiya, S., Kersting, A. B. & Ewing, R. C. [2009]. Groundwater nanoparticles in the far-field at the Nevada test site: Mechanism for radionuclide transport, *Environ. Sci. Technol.* 43: 1293–1298.

[77] Veerapaneni, S., Wan, J. & Tokunaga, T. [2000]. Motion of particles in film flow, *Environ. Sci. Technol.* 34: 2465–2471.

[78] Wan, J. M. & Tokunaga, T. K. [1997]. Film straining of colloids in unsaturated porous media: conceptual model and experimental testing, *Environ. Sci. Technol.* 31: 2413–2420.

[79] Wan, J. M. & Tokunaga, T. K. [2002]. Partitioning of clay colloids at air-water interfaces, *J. Colloid Interface Sci.* 247: 54–61.

[80] Wan, J. M. & Wilson, J. L. [1994]. Visualization of the role of the gas-water interface on the fate and transport of colloids in porous media, *Water Resour. Res.* 30(1): 11–23.

[81] Wan, J. & Tokunaga, T. K. [2005]. Comments on "pore-scale visualization of colloid transport and retention in partly saturated porous media", *Vadose Zone J.* 4: 954–956.

[82] Wan, J., Tokunaga, T. K., Kim, Y., Wang, Z., Lanzirotti, A., Saiz, E. & Serne, R. J. [2008]. Effect of saline waste solution infiltration rates on uranium retention and spatial distribution in Hanford sediments, *Environ. Sci. Technol.* 42: 1973–1978.

[83] Williams, D. F. & Berg, J. C. [1992]. The aggregation of colloidal particles at the air-water interface, *J. Colloid Interface Sci.* 152: 218–229.

[84] Xu, S., Gao, B. & Saiers, J. E. [2006]. Straining of colloidal particles in saturated porous media, *Water Resour. Res.* 42: W12S16, doi:10.1029/2006WR004948.

[85] Xu, S., Liao, Q. & Saiers, J. E. [2008]. Straining of nonspherical colloids in saturated porous media, *Environ. Sci. Technol.* 42: 771–778.

[86] Xu, S. & Saiers, J. E. [2009]. Colloid straining within water-saturated porous media: Effects of colloid size nonuniformity, *Water Resour. Res.* 45: W05501, doi:10.1029/2008WR007258.

[87] Zevi, Y., Dathe, A., Gao, B., Richards, B. & Steenhuis, T. [2006]. Quantifying colloid retention in partially saturated porous media, *Water Resour. Res.* 42: W12S03, doi:10.1029/2006WR004929.

[88] Zevi, Y., Dathe, A., McCarthy, J. F., Richards, B. K. & Steenhuis, T. S. [2005]. Distribution of colloid particles onto interfaces in partially saturated sand, *Environ. Sci. Technol.* 39: 7055–7064.

[89] Zhuang, J., McCarthy, J. F., Tyner, J. S., Perfect, E. & Flury, M. [2007]. In situ colloid mobilization in Hanford sediments under unsaturated transient flow conditions: Effect of irrigation pattern, *Environ. Sci. Technol.* 41: 3199–3204.

[90] Zimon, A. D. [1969]. *Adhesion of Dust and Powder*, Plenum Press, New York, NY.

Chapter 5

Seismic Safety Analysis and Upgrading of Operating Nuclear Power Plants

Tamás János Katona

Additional information is available at the end of the chapter

1. Introduction

1.1. Background

Experience of the March 11 2011, Great Tohoku earthquake clearly demonstrated that the earthquakes might be the dominating contributors to the overall risk of nuclear power plants (Institute of Nuclear Power Operations [INPO], 2011); International Atomic Energy Agency [IAEA], 2007). The seismic probabilistic safety assessments of several nuclear power plants also provided similar results. On the other hand, experiences show that plants survive much larger earthquakes than those considered in the design base, as it was the case of Kashiwazaki-Kariwa plant, where the safety classified structures, systems and components survived the Niigata-Chuetsu-Oki earthquake in 2007 without damage and loss of function (IAEA, 2007). In spite of the nuclear catastrophe of the Fukushima Daiichi plant caused by the tsunami after Great Tohoku earthquake 11th of March 2011, the behaviour of thirteen nuclear unit in the impacted area on the East-shore of the Honshu Island demonstrated high earthquake resistance. Consequently, proper understanding and assessment of the safety for the case earthquake (and generally for the external hazards) is very important for the operating nuclear power plants.

For the operating plants basic questions to be answered are, whether the nuclear power plant (NPP) is safe enough within the design basis and whether the operation can be continued safely if an earthquake hits the plant.

The designer and operators were mainly focusing on the first question, i.e. whether the reactor can be shut down, cooled-down, the residual heat can be removed from the core and spent-fuel stored at the plant, and the radioactive releases can be limited below the acceptable level in case of an earthquake. The second question became important especially after series of events when large nuclear capacities were shutdown for assessment of plant post-earthquake condition and justification of safety before their restart (Onagawa NPP in 2005, Shika NPP in 2007, Kashiwazaki-Kariwa NPP in 2007, Hamaoka NPP in 2009). Obviously, there is a need for reliable justification of plant safe status after felt earthquake for avoiding long shutdown time and consequent economic losses. Recently, the importance of the rapid assessment of the post-event plant status became very important from the point of view of the emergency management. This is one of the lessons learnt from the stress tests of nuclear operators following the Fukushima Dai-ichi accident.

Operators of nuclear power plants worldwide performed seismic re-evaluation and upgrading programmes of nuclear power plants during last three decades. A summary of international effort is given e.g. in (Campbell, et al, 1988; Gürpinar & Godoy, 1998) and in the special report issued by Nuclear Energy Agency, thereafter NEA, (NEA, 1998). The re-evaluation and upgrading of the seismic safety of the operating nuclear power plants were motivated mainly by the changing understanding of the seismic hazard at plant sites and/or recognition of inadequacy of design and/or qualification of certain safety related systems, structures and components relative to the seismic hazard or state-of-the-art of the technique and the requirements. In some countries, the existence of necessary margins with respect to the beyond design base earthquakes and avoidance of the cliff edge effects have to be demonstrated. The scope of the seismic re-qualification and upgrading programmes includes the definition of the pre-earthquake preparedness and post-earthquake actions at the plants.

All operating nuclear power plants in the United States are conducting an Individual Plant Examination of External Events, including earthquakes beyond the design basis, and about two-thirds of the operating plants are conducting parallel programs for verifying the seismic adequacy of equipment for the design basis earthquake; see (Campbell, et al, 1988). Western European countries also performed some re-evaluation of their older nuclear power plants for seismic events (NEA, 1998). Most extensive programmes have been performed in Eastern-European countries, where the operators implemented comprehensive programmes for evaluating and upgrading the seismic safety of their operating nuclear power plants (Gürpinar and Godoy, 1998; IAEA, 1995, 2000). Seismic re-qualification has been performed at following VVER plants:

-    NPP Paks, Hungary VVER 440-213, 4 units
-    NPP Mochovce, Slovakia VVER 440-213, 2 units
-    NPP V2 Bohunice, Slovakia VVER 440-213, 2 units
-    NPP Dukovany, Czech Republic VVER 440-213, 4 units
-    NPP Medzamor, Armenia, VVER 440(specific design) 1 unit
-    NPP Temelin, Czech Republic, VVER-1000, 2 units
-    NPP Kozloduy Unit 5 and 6, Bulgaria, VVER-1000, 2 units

The scope of seismic safety programmes at VVER-440/213 plants was the most extensive. It includes the re-evaluation of the hazard, reinforcement of structures and components, qualification of the active equipment, installation of seismic instrumentation and development of appropriate procedures. The seismic safety programme implemented at

Paks NPP was one of the most complex one. The implementation of measures was completed in 2003. Therefore the peculiarities of the programme, its scope and the applied methodologies could not be properly addressed and interpreted in the referenced above review papers. Originally the Paks NPP has not been designed and qualified for the earthquake loads. The seismic safety programme at Paks NPP has therefore aimed at design basis reconstitution. The re-evaluation of site seismic hazard included all required geological, geophysical, seismological and geotechnical investigations. The seismic design basis had been newly defined. Formally the compliance with design basis requirements has to be ensured by design methods and standards. It was already recognised that a consequent and full scope re-design in line with design codes and standards and subsequent upgrading might be impossible at Paks NPP. It should be recognised that use of methodologies developed for the justification of the seismic safety of operating plants does not ensure the compliance with design basis requirements and cannot be directly applied for VVER plants. The qualification of the nuclear power plant have been executed for the newly defined design basis earthquake by applying procedures and criteria for the design, combined with the methods and techniques developed for seismic re-evaluation of operating nuclear power plants. The selection and use of methodologies has been graded in accordance with safety and seismic classification of the SSCs. After implementing the measures for design basis reconstitution, the achieved level of safety has been quantified via seismic PSA, which provides the core damage frequency.

The question of the safe continuation of operation became very important as the World largest Kashiwazaki-Kariwa plant was shutdown for long-term after Niigata-Chuetsu-Oki earthquake in 2007 that caused a 0.67g ground motion acceleration at the site (value measured at the Unit 1 base mat). The safety classified SSCs designed for PGA 0.27g survived the earthquake without damage and loss of function while the non-safety structures were heavily damaged. The justification of the safety took two years.

The decision on the continuation of the operation is rather simple if the earthquake does not exceed the operational base (OBE) level. The case becomes more difficult if the OBE-level is exceeded and there are obvious damages in place. The justification of operability is even more complex if the earthquake loads exceed the design base level and there are no obvious failures/damages as it happened at the Kashiwazaki-Kariwa plant. Obviously, the judgement on the continuation of operation should be based on the set of information regarding capability of SSCs to survive an earthquake and on the post-event inspections, tests and analyses. It would be very reasonable to have in advance an assessment method for the plant status to ensure the effectiveness of the post-earthquake walk-downs and other actions, and to limit the time of shutdown. The methods for judgement on the safe continuation of operation can be developed on the basis of the design information. The results of the seismic probabilistic safety analysis (seismic PSA) or margin assessment provide useful additional information regarding weak-links. The design provides deterministic type information that no failure or damage should be expected if the earthquake loads do not exceed the design base level. However, the probability of damage is not zero even if the loads are less than the design base one. The seismic PSA provides the core damage frequency as the output of the analysis, which is a measure of the seismic safety. The PSA is generally failure oriented. The seismic PSA shows the weak links. This knowledge can be very useful for the planning of the post-event inspections. Similar information is provided by the seismic margin analysis, which quantifies the capability of the plant to survive an event greater than the design basis one.

After the severe accident at Fukushima NPP the operators in European Union, U.S. and some other countries including Japan performed comprehensive safety and risk evaluation of operating nuclear power plants, see e.g. (European Commission, 2011). These tests/reviews will launch different re-evaluation and upgrading programmes with regard to seismic safety and for improvement of the capability to cope with the beyond design base earthquake and associated events (fire, flood) at the existing plants. This process includes at some sites the re-assessment of the site hazard motivated by recent events and/or new scientific evidences, for example in the U.S. (NRC, 2011). These lessons learned will also affect the projects under preparation and/or implementation. For the new plants, it has to be demonstrated that the plant has sufficient margins with respect to the design basis extension earthquake loads of and avoiding the cliff-edge effect. Consequently, the lessons learned from the former projects for evaluation of the seismic safety and upgrading of operating plants are still of great practical importance.

1.2. Objective of the Chapter

Objective of the recent Chapter is to provide practical insights to the re-evaluation and upgrading of seismic safety of operating plants.

Evaluation of seismic safety and re-qualification of operating plants require specific approach; the safety goals have to be ensured in reasonable manner, avoiding unnecessary conservatism, contrary to the design that is ab'ovo conservative. State-of-the-art methodologies have to be implemented in every aspect of the re-evaluation and upgrading process. The optimisation of the measures from logistics point of view is very important under the condition of an operating plant.

International Atomic Energy Agency developed a comprehensive Safety Guide on "Evaluation of Seismic Safety for Existing Nuclear Installations" (IAEA, 2009a). The supporting document of this Guide is the Safety Report Series No 28 on "Seismic Evaluation of Existing Nuclear Power Plants" (IAEA, 2003) that summarises the before 2003 experience in the seismic evaluation and upgrading of the operating plants. These documents focus mainly on the methodologies for seismic safety evaluation that do not involve a change in the design basis earthquake.

In this Chapter the case of seismic evaluation and upgrading methodologies and solutions are presented. The Chapter includes the case for upgrading of an operating nuclear power plant originally not designed for earthquake. Based on the graded approach, the feasibility of the application of seismic design methods combined with those developed for the re-evaluation of existing plants is demonstrated.

New areas of the seismic safety evaluation of operating plants are also addressed in the Chapter that were triggered by recent events, the Kashiwazaki-Kariwa plant and the Fukushima Dai-ichi plant, that are focusing on the assessment and assurance of the beyond design base capability of the nuclear power plants, periodic review of safety, etc.

1.3. Scope of the Chapter

Scope of the Chapter covers

- the basic principles for ensuring the seismic safety of nuclear power plants,
- typical cases of the re-evaluation and re-qualification programmes, including cases of design basis reconstitution and studies for restart after an earthquake as well as the evaluation of the beyond design base capabilities of the plants,
- the applicable re-evaluation methodologies,
- the most important aspects of the pre-earthquake preparedness and post-earthquake actions,
- the full scope implementation example,
- the aspects of maintaining the seismic qualification during operation and periodic safety review.

1.4. Structure of the Chapter

Section 2 of this Chapter defines the basic principles of seismic safety. Section 3 provides an overview of the methodologies applicable: Section 3.1 outlines the objective and scope of the seismic safety programmes. Section 3.2 provides an overview of applicable methodologies. Sections 3.3 address the issues of restart after earthquake. Section 3.4 outlines the questions of accident management. Sections 3.5 to 3.7 address the walk-down, design of upgrading and role of the peer-review. Section 4 is devoted to the pre-earthquake preparedness and post-earthquake actions. The practical and full scope example of seismic re-evaluation and upgrading is shown in Section 5. Section 6 and 7 are related to the maintenance of the seismic qualification during operation and periodic safety reviews. Extensive list of references is provided to the Chapter in Section 8.

2. Basic principles of seismic safety

The fundamental safety objective of design and operation of nuclear power plant is to protect human life and environment in case of any malfunctions, failures of the plant systems, structures and components which may occur during the plant lifetime including those caused by rarely occurring earthquakes. The generic approach for ensuring this safety objective is the application of the concept of the defence in depth. In accordance with this concept, the following requirements are applicable:

1. Inherent and/or engineered safety features, safety systems and procedures have to be in place for the case of earthquakes
    a. for leading the plant to a safe shutdown state, i.e.
        -   for the maintaining the sub-criticality in the reactor and spent fuel pool and,
        -   for the cool down and heat removal from the core and spent fuel;
    b. for maintaining at least one protection barrier to ensure that the radiological consequences would be below the required limits.
2. Means, plans and procedures have to be in place for on-site and off-site emergency response to mitigate the consequences of accidents that result from failure of safety features and accident management measures in case of severe earthquakes.

The seismic safety is ensured by the following complex activities:

1. Site investigations and evaluation of the site seismic hazard, including hazards caused by the earthquake, like soil liquefaction;
2. Definition of the characteristics of the design basis earthquake;
3. Adequate design;
4. Use of qualified components;
5. Installation of seismic instrumentation;
6. Development of accident-prevention and accident-management procedures;
7. Evaluation of safety;
8. Seismic housekeeping;
9. Periodic safety assessment and subsequent upgrading if needed.

The basic safety functions, i.e. shut down, cooling and containment, have to be maintained for the earthquakes within the design basis envelope and with some extent for the severe beyond design basis earthquakes.

Traditionally the design of the nuclear facilities adapted the two-level concept: design for safety, using a high-level seismic excitation for design basis and design for production, using a moderate level of seismic excitation for operational limit.

The design base earthquake has to be defined with quite low probability of exceedance during operating time. This earthquake is the Safe Shutdown Earthquake (SSE) as per U.S. terminology; see U.S. NRC 10CFR Part 50, Domestic Licensing of Production and Utilization Facilities (NRC, 1956). It is called *Sicherheitserdbeben*, i.e. safety earthquake in German Nuclear Safety Standards 2101 (Kerntechnische Ausschuss [KTA], 1990), it is the maximum design earthquake (MRZ) according to the Russian-Soviet terminology and it is called SL-2 earthquake level by the IAEA guideline NS-G-1.6 (IAEA, 2003b) [4]. Here the term of Design Base Earthquake (DBE) will be used. According to the international practice the annual probability of exceedance of the DBE is usually 10^{-4}/year in case of nuclear power plants. The lower limit of the peak ground acceleration (PGA) of the DBE is set for 0.1g regardless of the site (article 2.7 of NS-G-1.6). The shutdown and cool-down of the reactor, the continuous heat removal from the irradiated fuel (in the reactor core and spent fuel pool), and the limitation of releases have to be ensured in this limit state. SSCs required for basic safety function have to sustain the earthquake loads without loss of function.

Operability of NPPs should be ensured after the moderately frequent and not severe earthquakes. The operational base earthquake (OBE or SL-1 level according to the IAEA terminology) level is defined as a design level for continuous operation. The OBE was usually defined as an event with frequency of 10^{-2}/a, or a ground motion with maximum horizontal acceleration equal to a given fraction of the maximum acceleration value of the SSE. Through the years the concept of designing for two earthquakes has radically changed. Nowadays, the OBE is interpreted as an operational limit and inspection level rather than an obligatory design level. The definition of the OBE level is subject of design, operational, economic considerations; see the IAEA NS-G-3.3 Safety Guide (IAEA, 2002). Design for lower level is not required if the OBE PGA is equal or less than 1/3rd of the SSE PGA, see in Appendix S of the 10 CFR Part 50. Instead of OBE PGA, new criteria for the exceedance of operational limit/inspection level are introduced. The changes of the terminology in the German regulation demonstrate the changes in design concept: the former terms SSE - *Sicherheitserdbeben* and OBE - *Auslegungserdbeben* were replaced by the terms design base earthquake and inspections earthquake, i.e. *Bemessungserdbeben* and *Inspektionserdbeben*.

3. Tasks for seismic re-evaluation and upgrading of operating NPPs

Major tasks of the seismic re-evaluation and upgrading projects are

- identification of the objective and scope of the programme
- selection of the methods for the re-evaluation, including definition of the seismic input for the analyses and performance of the analyses
- design/development and implementation of modifications and re-qualification measures
- evaluation of the achieved safety level, calculation of the core damage frequency due to earthquake.

The tasks are determined by the objective of the project as it has been shown above, i.e. resolution of qualification issues, ensuring the design basis compliance, etc.

3.1. Objective and scope of the seismic safety programmes

Generic objective of the seismic safety programmes is to ensure the basic nuclear safety functions, i.e.

- the control of the reactivity in the reactor and spent fuel pool, i.e. the ability to shutdown the reactor and maintain the sub-criticality after the earthquake,
- to cool down and heat removal from the core and spent fuel,
- to maintain the containment function for the reactor and spent fuel, i.e. limit the release of radioactive substances into the environment.

The functions have to be maintained for the earthquakes within the design basis envelope and with some extent for the earthquakes with parameters exceeding the design basis one.

The basic concept of the seismic safety re-evaluation and of the operating nuclear power plants, and the selection of the methods and criteria is different from that are required in case of the design of new power plants; see the *INSAG-8* document *"A Common Basis for Judging the Safety of Nuclear Power Plants Built to Earlier Standards"* (IAEA, 1995).

The graded approach is used while ensuring the seismic safety of NPPs, i.e. the safety importance of the SSCs is considered and according to this the SSCs are classified into seismic safety classes, which define the requirements assigned to the design, qualification and operation of the SSCs. Well-defined set of plant systems and structures and components are required to be functional during and after the earthquake for bringing the plant in-to stable shutdown condition. Some of those SSCs are passive, e.g. the pressure retaining boundaries or the containment. They shall sustain the vibratory load remaining leak-tight; however some plastic deformation, ductile behaviour might be allowed. In some cases the deformation has to be limited to the elastic for ensuring some active functions. Building structures and equipment supporting structures might be also loaded to plastic region up-to the level, which does not impair the intended safety functions. The active systems functionality requires qualification for the vibratory motion as well as availability of supporting functions, e.g. electrical power supply.

Practically, a conscious and careful evaluation and utilisation of the built-in margins provide the possibility for achieving the target safety level at operating plants by feasible amount of modifications and re-qualifications.

The scope and the methodology of the seismic safety programmes vary with the motivation of the particular project. Practically there have been three different objectives of the past seismic safety programmes:

1. to resolve the inadequacy of the design and qualification while the seismic design basis remains unchanged, i.e. to comply with the current licensing basis;
2. to comply with newly defined seismic design basis (modification of design basis either because of new scientific evidences regarding seismic hazard or because of changing regulations);
3. to evaluate and demonstrate the seismic margin.

The objective and scope of recent seismic safety re-evaluation programmes is to demonstrate the plant safety for the design base extension, to justify the re-start after strong earthquake, and to identify the plant vulnerability in case of severe event and develop adequate accident management provisions.

3.1.1. Resolving the inadequacy issues

Example for the first type of seismic safety programme is the resolution of USI A-46 seismic issues of older, operating nuclear power plants in the U.S. (NRC, 1987). This programme was aimed to demonstrate the seismic adequacy of essential equipment at older operating plants by the use of available seismic experience data for similar equipment. The rules for the resolution of the USI A-46 issues are defined in the Generic Implementation Procedure, thereafter GIP, developed for Seismic Qualification Utility Group [SQUG] (SQUG, 1992). The scope of the programme was limited to the equipment needed for the safe shutdown of the reactor after a design basis earthquake and bringing the plant to a stable hot or cold shutdown condition for as minimum 72 hours of time. A single shutdown path and a backup for decay heat removal were defined. The seismic input used for the qualification was set to the SSE and the design floor response spectra. The core of the GIP is the empirical qualification method and database. The GIP was applied in several countries, e.g. U.K. and Belgium.

3.1.2. Seismic margin programmes

It is important to demonstrate on one hand that the nuclear power plant will remain safe in case of an earthquake that exceeds the design base level, whether the basic safety functions can be lost due to sudden failure (i.e. 'cliff-edge' effect). On the other hand it is important to know the contribution of the seismic hazard to the plant core damage frequency. Example for margin assessment and quantification of the seismic safety in terms of core damage frequency is the NRC initiated Individual Plant Examination of External Events, thereafter IPEEE in the U.S. (NRC, 1991). There are three methods for the margin assessment: the seismic PSA, and margin assessment using either the deterministic method developed by EPRI or the probabilistic method developed by the NRC. In this case of deterministic method a reference level earthquake is selected for which – under certain assumptions – the capacity has to be demonstrated. The scope of SSCs considered in the margin assessment depends on the method selected, e.g. in case of seismic PSA the scope of SSCs is identical to the Level 1 PSA plus the containment.

3.1.3. Reconstituting the design basis

The most demanding programmes were those for ensuring the compliance with newly defined design basis.

These programmes include the following tasks:

1. Evaluation of the seismic hazard of the site that includes the associated with earthquake events, e.g. liquefaction;
2. Development of the design basis earthquake characteristics;
3. Identification of the structures, systems and equipment, which are needed for ensuring that basic safety functions;
4. Evaluation of the seismic capacity of SSCs and identification of the upgrading;
5. Design and implementation of the necessary corrective measures;
6. Installation of seismic instrumentation;
7. Development of pre-earthquake preparedness and post-earthquake measures;
8. Evaluation of the safety, i.e. quantification of the core damage frequency due to earthquake, quantification of the safety margins.

Depending on the case and the national regulation, the scope of the design base reconstitution programme can cover either all SSCs classified into seismic and safety classes as per new design, or the scope is limited to the SSCs required for safe shutdown and bringing the reactor into stable (hot or cold) shutdown condition. Those non-safety classified SSCs have to be also considered damage/failure of which can disable certain safety functions due to seismic interactions (falling down, flooding, fire).

3.1.4. Recent beyond design base studies

The quantification of the margins has three aspects:

- it is part of the design,
- it is needed while evaluating the plant condition and justifying the restart after a strong earthquake hit the plant,
- it is needed for the development of the severe accident management provisions.

According to the IAEA design requirements NS-R-1 (IAEA, 2000), *the seismic design of the plant shall provide for a sufficient safety margin to protect against seismic events.* This means that the abrupt lost of function has to be excluded by the design even if the earthquake demand exceed the design base one (see also NS-R-1.6 paragraph 2.39 regarding 'cliff-edge' effect).

According to the novel requirements, the capability of the new plants to withstand the loads and conditions of the design basis extension has to be ensured by the design provisions. In case of new plants, a minimum configuration of SSCs for ensuring the shutdown and subcriticality of the reactor, heat removal to the ultimate heat sink and the containment have to remain functional for the accident management purposes. A margin type evaluation has to be performed for demonstration the beyond design base capabilities of the new plants (1.4 times the SSE loads as per EUR requirements and 1.67 times of the SSE loads in the U.S. practice). Best estimate methods can be used for the justification beyond design base capabilities.

The plant safety re-assessment after a strong earthquake requires an overall checking the post-event condition of all SSCs, even those non-safety classified SSCs, since both the safety and operability have to be demonstrated. The possible analysis and testing/inspection methods should be selected and applied in accordance with safety relevance and impact on the operation (Nomoto, 2000). According to (Kassawara, 2008), the probabilistic margin analysis can also be effective in this case.

Recently, the availability of severe accident management provisions become of great importance. The scope of stress tests covers review of compliance with design base requirements, demonstration of beyond design base capacity (avoidance of the cliff-edge effect) and identification of plant vulnerability/damage state and development of severe accident management measures and guidelines. Generally, some margin type analyses have been performed in the participating countries for the possible minimum configurations needed for shutdown and heat removal of the reactor and spent fuel and protection of the containment. Identification of seismic interactions (fires, flooding, logistical obstacles) became important since these can affect the function of the SSCs within the minimum configuration, inhibit the connections of provisory power and cooling lines, impeding the implementation of Severe Accident Management/mitigation measures as it is to see in the country reports at the European Nuclear Safety Regulators Group site (ENSREG, 2012).

3.2. Methodologies for re-evaluation of seismic safety

The methodologies for the seismic re-evaluation and re-qualification are as follows:

1. Qualification by empirical methods
2. Quantification of margins:
    a. Seismic Probabilistic Safety Assessment (NRC, 1983)
    b. NRC Seismic Margins Method (Budnitz et al., 1985; Prassinos et al., 1986)
    c. Electric Power Research Institute Seismic Margins Method (EPRI, 1988)
3. Design methods – justification by analysis

3.2.1. Qualification by empirical method

Empirical qualification of the plant equipment is a powerful tool for seismic re-qualification of operating NPPs. The empirical qualification methods have been recognised by IAEA in the Safety Guides NS-G-1.6 as well as NS-G-2.13.

The empirical qualification database developed for SQUG covers twenty classes of equipment, e.g. active equipment as well as cable raceways, tanks and heat exchangers (SQUG, 1992; Starck&Thomas, 1990), except of pipelines and structures. As an alternative solution, the U.S. Department of Energy [DoE] has developed the Seismic Evaluation Procedures, a procedure similar to GIP that also covers pipelines and ventilation ducts (DoE, 1997).

The steps of the Generic Implementation Procedure are as follows (SQUG, 1992):

-   Development of safe shutdown equipment list
-   Development of seismic demand (in-structure response)
-   Equipment walk-down and screening
-   Relay evaluation
-   Outlier resolution
-   Reporting

The methodology and the database (the so called SQUG-database) can be adapted to the needs of different programmes for the resolution of design/qualification inadequacy issues.

Generally the process has to be started with development of the list of SSCs requiring re-qualification for a given level of earthquake. The basis of the identification of the scope can be the list of SSCs for safe shutdown or the seismic and/or safety classification database as it was the case at Paks NPP.

Four criteria are used for the verification of seismic capacity: (1) Comparison of the seismic demand to the SQUG bounding spectrum; (2) Checking in the experience database (caveats and inclusion rules); (3) Checking the anchorage; (4) Evaluation of the seismic interactions.

The seismic demand can be defined either by design floor response spectra, or by scaling-up the design floor response spectra to the required level, or by completely new response calculation for the required input (e.g. at Paks NPP the floor response spectra have been calculated for the newly defined DBE).

In case of most of this equipment, the load-bearing capacity is verified by demonstrating that the equipment is adequately anchored. Operability is demonstrated by verifying that the equipment is similar to the equipment of the database created on the basis of experience and that it meets all of the prescriptions included in the GIP.

Important element of the procedure is the walk-down that provides the basis for screening out the obviously rugged items and for the consideration of the as built conditions, since in majority of cases, the load-bearing capacity is ensured, if the equipment is adequately anchored.

The applicability of the empirical qualification method should be carefully checked via reviewing the similarities between the features of the items in the database and item to be qualified at the plant. The empirical method and database were adapted for the qualification of the VVER equipment (Masopust, 2003) and used for the qualification of the VVER at Paks NPP Hungary, Bochunice and Mochovce NPP in Slovakia, though the objective and scope of the particular seismic safety programmes differed very much from those in the USI A-46.

3.2.2. Deterministic margin analysis

The plant design shall ensure sufficient margins against seismic demand, as it is required by the IAEA design requirements NS-R-1 (IAEA, 2000).

In case of operating plants, the objective of the seismic margin assessment (SMA) is to evaluate, quantify the inbuilt seismic margin of those structures, systems and components of the power plant that fulfil their basic safety functions during and after the earthquake. The quantification of the margins is also recommended by the IAEA Safety Guides NS-G-1.6 and NS-G-2.13 (IAEA, 2003 and 2009). The goal of the analysis is to determine the seismic shaking level at which there is a high-confidence-of-low-probability-of-failure (HCLPF). This HCLPF is mathematically defined as 95% confidence of less than 5% probability of failure.

In SMA calculation the seismic capacity C_S is to compare to the Seismic Margin Earthquake (SME) demand D_S. The capacity and the demand have to be calculated according to codes and standards while some specific assumptions should be accepted. These assumptions are as follows (EPRI, 1998):

- Load combination has to consist of normal and SME loads. The ground response spectrum is median-shaped.
- Conservative estimate of median damping have to be used.
- Best estimate structural model has to be used and the uncertainty in frequencies has to be accounted.
- Calculation of the soil-structure interaction has to be best estimate taking into account the parameter variation.
- Code specific minimum strength or 95% non-exceedance probability values
- Static capacity equations to be used have to be for code ultimate strength (ACI) for concrete, or maximum strength, (AISC) for steel structures, or Service level D (ASME) or functional limits in case of mechanical equipment.

- Inelastic energy absorption values to be used for non-brittle failure modes and linear analysis can be taken e.g. from (IAEA, 2003a)
- In-structure spectra have to be calculated by frequency shifting rather than peak broadening to account for uncertainty while median damping is used.

The capacity-demand ratio for elastic response $(C/D)_E$ is:

$$\left(\frac{C}{D}\right)_E = \frac{C - \Delta C_S}{D_S + D_{NS}} \tag{1}$$

where D_{NS} is the concurrent non-seismic demand for all non-seismic loads in the load combination, ΔC_S is the reduction of the capacity due to concurrent seismic loading. The inelastic capacity-demand ratio $(C/D)_I$ can be similarly calculated taking into account the ductility F_μ. If the inelastic capacity-demand ratio $(C/D)_I$ exceeds unity the seismic margin earthquake level SME exceeds the reference level earthquake RLE for what the existence of sufficient margin has to be demonstrated. Otherwise, the built-in capacity that can be utilized for sustaining the seismic demand is equal to $C_S = (C - D_{NS})$, the seismic demand is equal to $(D_S + \Delta C_S)$. The RLE (or more precisely the PGA of the RLE) has to be scaled by the ratio $((C - D_{NS})/(D_{NS} + \Delta C_S))$ in elastic response case; or by $F_\mu((C - D_{NS})/(D_{NS} + \Delta C_S))$ when for inelastic response considered. That value will be the code deterministic failure margin with high confidence for low probability of failure (HCLPF) expressed in terms of the peak ground acceleration, i.e.

$$HCLPF = \frac{C - D_{NS}}{D_S + \Delta C_S} F_\mu a_{RLE}. \tag{2}$$

The seismic capability active equipment (electrical, electromechanical and I&C) is qualified by tests or empirical method (see Section 3.2.1 above). Based on the qualification or fragility test data or generic data, a bound of the test response spectra have to be defined at about the 99 per-cent exceedance probability level. The in-structure response spectra calculated for the reference level earthquake and the ratio of the bound of the test response spectra and in-structure/floor response spectra has to be calculated. Scaling up the reference level (PGA) with this ratio provides the HCLPF capacity of the equipment. The HCLPF capacity that has to be evaluated for all items needed for ensuring the basic safety function.

A systematic SMA procedure and methodology has been developed by EPRI in NP 6041 Rev 1 consisting of following main elements (EPRI, 1998):

1. Definition of the Review Level Earthquake (RLE)
2. Identification of success paths needed to bring the reactor into stable
   a. Two independent functional paths to shutdown
   b. Define components in the paths
3. Plant walk-down
   c. Screening out the rugged components
   d. Identify characteristics, vulnerability
   e. Assures verification of as-installed properties and conditions
   f. Identify interactions
4. Seismic capacity evaluation for unscreened components

The SMA seismic input is the Review Level Earthquake (RLE) that should exceed the SSE. The RLE is that screening level at which structures, systems and components, necessary for the shutdown of power plant and for keeping it in the stable shutdown condition and considered to be in the 'success path', should be examined. (According to the definition of given in EPRI NP-6041 report, SME is equivalent to RLE specified by NUREG-1407. There are three categories of sites according to the PGA: PGA≤0.3g, 0.3<PGA≤0.5g and PGA>0.5g with reference level PGA 0.3g, 0.5g and >0.5g respectively. For the analysis, the NUREG/CR-0098 median shape ground motion response spectral can be selected.)

In margin analysis, the success path selection must include a primary success path and an alternate success path utilizing to the greatest extent possible, different equipment. One of the paths must also have the capability to mitigate a small pipe break.

The rugged components have to be screened out during the plant walk-down. For those components that were not screened out during the walk-down phase, additional analyses shall be executed to determine the HCLPF. The weakest component in a shutdown path then defines the plant level HCLPF for that path.

For the new design, the margin beyond safe shutdown earthquake has to be demonstrated (HCLPF for at least 1.67 times of the SSE in the US design practice and 1.4 times of the SSE in the European design practice).

The seismic margin assessment procedure is experience and expert judgment driven. Therefore the selection of the team is a decisive precondition for success and adequacy of the result. The development of the safe shutdown equipment list and performance of the walk-downs require very experienced team consisting of systems engineer, structural/seismic engineers trained in design and empirical qualification as well.

3.2.3. Probabilistic seismic safety analysis (seismic PSA)

One of the most complex cases for assessing the nuclear power plant (NPP) safety is the evaluation of the response of the plant to an earthquake load and the risk related with this. The objective of the seismic PSA is to define the contribution of the earthquakes to the core damage frequency of the reactor and finally to the overall risk of plant operation. Risk is expressed as triplets $R = \{\langle S_i | p_i | L_i \rangle\}$, where S_i is the identification/description of the ith scenario or accident sequence; p_i is the probability of occurrence of that scenario and L_i is the measure of the consequences/losses caused by that scenario.

In case of earthquake, the probability of damage/failure of a structure or component P_{fail} depends on a rather complex load vector $X = (x_1, \ x_2, \ ...)$ that expresses all features of the earthquake excitation (peak ground acceleration, duration of strong motion and frequency distribution of the energy of excitation). The P_{fail} can be calculated as follows:

$$P_{fail} = \int_R \ h(x_1, \ x_2, \ ...) \, P(x_1, \ x_2, \ ...) dx_1 dx_2 \, ..., \tag{3}$$

where the $h(x_1, x_2, \dots)$ represents the hazard, i.e. it is the probability density function of applied loads and $P(x_1, x_2, \dots)$ denotes the conditional probability of failure. The Equation (3) is theoretically precise. Nevertheless, in the practice the peak ground acceleration is used as a single load parameter. There were also some attempts made for using the cumulative absolute velocity for load parameter (Katona, 2010, 2011).

The basics of the seismic PSA were outlined in (Kennedy & Ravindra, 1984). Frequencies of core damage caused by an earthquake are calculated by modelling of the plant behaviour by event trees constructed to simulate the plant system response. Fault trees are needed for the development of the probability of failure of particular components taking into account all failure modes. The hazard is expressed as complementary probability: 1-cumulative probability function, i.e. probability that the peak ground acceleration exceeds a given value. The fragility is defined as the conditional probability of core damage as a function of a – the PGA at free surface. The behaviour of the plant is modelled by the Boolean description of sequences leading to failure. Plant level fragility is obtained by combining component fragilities according to the Boolean-expression of the sequence leading to core damage. The plant level fragility is defined as the conditional probability of core damage as a function of free field PGA at the site. Plant level fragilities are convolved with the seismic hazard curves to obtain a set of doublets for the plant damage state. A great number of studies have been published on the seismic PSA, a review and referencing all of them is impossible in the frame of recent study. The method is now well developed and standardised by ASME in ASME/ANS RA-S–2008 (ASME, 2008) (see also the addendum ASME/ANS RA-Sa–2009).

According to ASME/ANS RA-S-2008, for evaluation of core damage frequency the doublets $\{\langle p_{ij} | f_{ij} \rangle\}$ have to be calculated, where f_{ij} is the frequency of the earthquake induced plant damage state,

$$f_{ij} = \int_0^\infty f_i(a') \frac{dH_j}{da} da', \tag{4}$$

where p_{ij} is the discrete probability of this frequency $p_{ij} = q_i p_j$; q_i is the probability associated with i^{th} fragility curve $f_i(a)$ and p_j is the probability associated with j^{th} hazard curve H_j. The seismic fragility $f_i(a)$ is the conditional probability of failure for a given value of seismic input parameter, e.g. peak ground acceleration. The fragility curve $f_i(a)$ is the ith representation of the conditional probability of the core damage. In the practice both the hazard and fragility is accounted by point estimates with subsequent uncertainty evaluation. The fragility is modelled by lognormal distribution:

$$f(a) = \int_0^a \frac{1}{\sqrt{2\pi}\beta_c x} e^{-\frac{(lnx-\mu)^2}{2\beta_c^2}} dx, \tag{5}$$

where $\mu = ln\,C_m$, is the logarithm of median capacity C_m, and the β_c is the composite logarithmic standard deviation expressing the epistemic and aleatory uncertainty. The lognormal distribution of the fragility is a consequence of the representation of the median capacity, C_m, as a product and large number of different factors, F_i representing the uncertainties of all contributing to the capacity factors as well as the uncertainty in demand, i.e. $\prod_i F_i$. According to the central limit theorem the sum of random variables tends to the normal random variable independent form the distribution of each of them. This rule is applicable to the logarithm of the product above.

The HCLPF is related to the C_m as follows:

$$C_m = HCLPF \cdot e^{(2.3264\beta_c)}.\qquad(6)$$

The $h(a) = \left(\frac{dH_j}{da}\right)$ is the probability density function of the applied seismic load expressed in terms of peak ground acceleration, taken from the jth hazard curve. The form of the hazard curve is as follows (McGuire et al., 2001):

$$h(a) = k_0(a)^{-k},\qquad(7)$$

where k_0 and k are constants that can be defined on the basis of probabilistic seismic hazard assessment (PSHA).

The basic steps of seismic PSA are the followings:

1. Determination of seismic hazard by PSHA
2. Systems analysis
    a. Fault trees and event trees
    b. Define accident sequences, associated systems, components
3. Fragility analysis of SSCs – Conditional failure probability
4. Integration of hazard and fragility resulting in seismic core damage frequency.

Since the level of core damage probability to be assessed is very low, the assessment of seismic hazard has to be performed up to very low level of annual probability, e.g. up-to 10^{-7}/a or less. The median hazard curve can be used which can be defined adapting the guidance in the IAEA Safety Guide SSG-9 (IAEA, 2010).

The consideration of uncertainty in both fragility and seismic hazard is important for adequate safety assessment. The above formulation uncouples the uncertainties in the load and resistance parameters, embodied in the in the fragility and load probability density functions respectively. These uncertainties are usually of different origins and it is convenient to be able to treat them separately.

The level 1+ seismic PSA gives estimation for the probability of seismic induced reactor core-melt. The level 1+ means the examination of containment that includes the evaluation of the safety of containment integrity and isolation as well as the development of bypass.

The experiences of the seismic probabilistic safety (risk) studies performed in the U.S. are summarised in (NRC, 2010). In comparison with core damage frequency (CDF) due to internal initiators, the seismic core damage frequencies seem to be dominating. Similar conclusion can be made regarding seismic PSA results obtained for the Paks NPP, where the contribution of the seismic events to the total CDF is approximately equal to 75 per-cent of the total CDF; see also (Riechner et al., 2008) on the Swiss experiences. Generally, the acceptable level of the annual probability of reactor core damage due to seismic events is of order of magnitude 10^{-5}. It seems that the uncertainties dominate the seismic CDF caused by both uncertainty of the hazard definition, especially in the range below 10^{-5}/a frequencies, and by the uncertainty of the fragilities.

An interval representation can be proposed for accounting the uncertainty of the fragility (Durga et al., 2009; Katona, 2010). The fragility is a doublet $\{\langle p_i | f_j \rangle\}$ composed from set of fragility functions $f_i(a)$ with probability weight q_i, where the variable a is the horizontal component of the ground motion acceleration. It can be represented by a probability box (p-box), $\{\langle p_i | f_j \rangle\} \rightarrow [\overline{F}(a), \underline{F}(a)]$, where $F(a)$ is the conditional probability distribution of the failure. The $[\overline{F}(a), \underline{F}(a)]$ is the probability-box specified by a left side $\overline{F}(a)$, and a right side $\underline{F}(a)$ distribution functions, where the relations $\overline{F}(a) \geq \underline{F}(a)$ and $\underline{F}(a) \leq F(a) \leq \overline{F}(a)$ are valid.

The most trivial case for the use of p-box can be the screening according to ruggedness of the component. The rugged components might be described by p-box with lower and upper bounding value of the variable a, or any other damage indicator, i.e. cumulative absolute velocity. The probability bounds can be defined via expert elicitation.

It can also be convenient to express the uncertainty of fragility in form of a p-box, defined by a lower bound $u(p)$ and an upper bound $d(p)$ on the function $L^{-1}(p)$ defined as inverse of the probability distribution $F(a)$, i.e. $d(p) \geq L^{-1}(p) \geq u(p)$, where p is probability level.

As it was mentioned above, in the practice the lognormal distribution is applied for fragility of structures. If the bounds on mean, μ and standard deviation σ of a lognormal distribution L are known, $\alpha \in \{(\mu, \sigma) | \mu \in [\mu_1, \mu_2], \sigma \in [\sigma_1, \sigma_2]\}$ the bounds on the distribution can be obtained by computing the envelope of all lognormal distributions L that have parameters within the specified intervals: $d(p) = \max_\alpha L_\alpha^{-1}(p)$ and $u(p) = \min_\alpha L_\alpha^{-1}(p)$.

3.2.4. Probabilistic margin analysis

The NRC seismic margins method (NUREG CR 5334) is a truncation of PSA, i.e. the plant systems are modelled by Boolean method, while the systems needed for ensuring the basic safety functions are considered, the seismic fragility curves are developed, and the plant level HCLPF is computed (Campbell, 1998). The procedure does not involve the use of a seismic hazard for the computation of the HCLPF and the core damage frequency is not calculated.

The NRC seismic margins method involves the following steps (Prassinos at al, 1986):

-   Selection of the review level earthquake

- Development of systems models
- Initial component ruggedness screening
- Plant walk-down
- Development of component and structural - fragilities
- System analysis
- Determination of plant level HCLPF

The systems models and fragility curves are used to determine the dominant accident sequences and the plant level HCLPF. The RLE selection and walk-down procedures are similar to those used in the EPRI margin method. The screening is conducted to eliminate many components from fragility computations.

The HCLPF capacities for components in each system included into the plant model have to be defined and combined according to the Boolean representation of the system via minimum-maximum procedure: minimum HCLPF of the elements connected by or-gate and maximum HCLPF of the elements connected by the and-gate. For example, if the Boolean representation of a system composed from elements A, B, C and D is equal to A*(B+C)*D then the HCLPF of the system is equal to *Maximum of (A; Minimum of (B, C); D)*. The plant HCLPF can also been calculated via convolution procedure.

3.2.5. Use of design methods and standards

A consequent use of design methods and standards for the re-evaluation and upgrading is not practicable for the operating plants. However, in case if the plant was not designed for earthquakes or the hazard was very underestimated, the design methods and standards have to be used for achieving the compliance with design basis requirements as much as practicable.

The graded approach has to be applied for appropriate selection of evaluation methods. Deviation from design procedures can be accepted in case of qualification of outliers of Class 3 (seismic classification see e.g. IAEA Safety Guide NS-G-1.6 (IAEA, 2003b), safety classification principles are given e.g. in the IAEA NS-R-1 (IAEA, 2000b).

The possibility of differentiation at design is exposed by guideline NS-G-1.6:

Class 1: design ensuring the function and great safety margin are necessary

Class 2: items are classified because of seismic interactions; they 'can be designed with smaller safety margin'

Class 3: these can be designed differentiated according to hazard

Class 4: general industrial standard can be used

The assumptions and methods applicable for each tasks of the seismic re-evaluation and upgrading of the operating plants with the aim of design basis reconstitution are given in the Table 1. Practical example is given in Section 5 below.

Task	Practicable/advisable method
Evaluation of the seismic hazard of the site that includes the associated with earthquake events, e.g. liquefaction; Development of the design basis earthquake characteristics;	As for new design, preferable PSHA (see the IAEA NS-R-3, SSG-9 and NS-G-3.6) DBE as for new design – The Ground Motion Response Spectra have to be modified in accordance to ASCE/SEI 43-05 (ASCE, 2005) and Reg. Guide 1.208 (NRC, 2007) to be taken for design basis response spectra.
Identification of the structures, systems and equipment, which are needed for ensuring that basic safety functions;	According to the safety and seismic classification plus interacting SSCs. Stable shutdown conditions have to be ensured as minimum for 72 hours. Single failure criterion has to be applied.
Evaluation of the seismic capacity of SSCs and identification of the upgrading;	Graded approach: Class 1-3 evaluation by analysis according to design codes; Class 1 and 2 outliers has to be fixed; For Class 3 outliers justification via less conservative method (realistic damping, inelastic response) or upgrade;
Design and implementation of the necessary corrective measures (fixes and qualifications);	Design of modification according to codes and standards. Qualification by tests or empirical method.
Installation of seismic instrumentation; Development of pre-earthquake preparedness and post-earthquake measures;	According to the IAEA NS-G-1.6, NRC Regulatory Guide and 1.12, 1.166 and 1.167, (IAEA, 1995; NRC, 1997a, 1997b, 2000, EPRI, 1988, 1989, ANS 2002)
Evaluation of the CDF due to earthquake, quantification of the safety margins.	Seismic PSA

Table 1. Assumptions and methods applicable while complying with newly defined design basis requirements

3.3. Studies for restart after strong earthquake

There are specific procedures developed for the evaluation of the plant safety after a strong earthquake that are part of the plant emergency procedures (EOPs) for the case of earthquake, e.g. (NRC, 1997a, 1997b), (EPRI, 1989), (ANS, 2002) and (IAEA 2011).

The post-earthquake evaluation is in principle eclectic. The practicable methods are e.g. the evaluation by analysis, margin assessment, checking the post-earthquake condition of equipment along empirical criteria, in-service inspections and testing. Selection of the method can be performed on the basis of walk-down and visual inspection's experiences, safety classification, etc. Lessons learnt from the case of the Kashiwazaki-Kariwa plant after the 2007 earthquake are of great importance.

The justification of the continuation of the operation after a strong earthquake (even if it is below the SSE-level) is a rather complex issue.

The design is success oriented. Consequently, the comparison design versus experienced parameters provides basis for a deterministic statement, whether an SSC will fail or not.

The seismic PSA is failure oriented, it indicate the week links that have to be carefully checked. The margin studies quantify the built-in capacities/reserves that may cover the demand even beyond the design base, see (Kassawara, 2008).

Although it is the most time-consuming and expensive, the careful testing and the implementation of state-of-the-art analysis methods and removing the unnecessary conservatism of material parameters (mainly the damping) seems to be the most powerful tool for the evaluation of post-event situation.

There is an obvious need for a better damage indicator as the PGA and response spectra of the experienced earthquake and the comparison of these to the design base PGA and response spectra. The cumulative absolute velocity (CAV) is a good indicator for no damage according to the EPRI study (EPRI, 1988). Some recent studies show that the CAV can be used for damage indicator for assessing the post-event conditions, especially for the fatigue failure mode, since the CAV can be correlated to the product of the number of load cycles and the stress amplitudes, thus the fatigue lifetime limit can be written as a function of the CAV (Katona, 2011). Comparing of the Niigata-Chuetsu-Oki earthquake in 2007 and the Great Tohoku earthquake in 2011, the most significant difference is not in the PGA but in the overall energy of the ground motion that is properly characterised by the CAV value.

3.4. Severe accident management oriented studies

Recently, the severe accident management (SAM) studies with regard to extreme environmental conditions and hazards become great importance. For the planning of the accident management and mitigation measures and development of the severe accident management guidelines, the possible accident scenarios have to be known and the plant vulnerabilities and robust features have to be identified. For the design of technical means for the accident management/mitigation, the post-event conditions have to be forecasted.

For the adequate preparation for severe accident situations, simultaneous occurrence of extremities has to be assumed. Occurrence of additional earthquake induced events has to be expected, if the beyond design basis hits the plant. For example soil liquefaction can be the dominating issue and cause cliff-edge effect on soft soil sites if a strong beyond design base earthquake hits the site, while the liquefaction may not happen in design base case.

It has to be also assumed that extreme conditions, including logistical obstacles due to on-site and off-site damages will be in place while the accident management measures have to be implemented.

The seismic PSA and margin type analyses provide the basis for the definition of the possible damage sequences and identification of effective measures. According to the PSA experience, the most serious is the sequence of the total loss of power and possibility of the heat removal to the ultimate heat sink or even loss the ultimate heat sink while the containment isolation is lost with or without of containment isolation with or without significant structural damage of the containment.

Essential task of the studies related to severe accident management is the aseismic design of the connections of the provisional systems for cooling the reactor and spent fuel pool (pipelines for cooling and DC/AC power cabling and connections). The design basis of these provisions has to be defined well beyond the plant "usual" design basis. The seismic hazard curve should be available for this reason.

The concept and the main tasks of the severe accident management studies are as follows:

1. identification of possible minimum configurations needed for shutdown and heat removal of the reactor and spent fuel and protection of the containment
2. identification of provisional and mobile tools for ensuring the heat removal and containment protection
3. plant walk-down for
   a. screening out the robust elements
   b. identification of interactions affecting the SSCs within the minimum configuration and the inhibiting the connections of provisional power and cooling lines
   c. identification of the logistical obstacles impeding the implementation of SAM measures
4. identification of the measures needed for SAM
5. assessment/quantification of the margins

3.5. Role of the walk-downs

As it has been shown in the Sections above, the walk-down of the power plant is a key element of the seismic re-evaluation and re-qualification of the operating NPPs. The walk-downs provide the opportunity to see what is difficult or impossible to recognise just looking, reviewing the documentation. The aim of the walk-down is as follows:

1. to check the as-is conditions, i.e.
   a. the as-is lay-out conditions,
   b. the adequacy of the anchorages,
   c. to check the compliance with the conditions in the re-qualification database,
2. to identify those interactions, which can potentially affect the performance of the seismic safety related structures, systems and components during the occurrence of an earthquake and can render this equipment inoperable,
3. to check the feasibility of upgrading measures.

Examples for checking the interaction items during the walk-down are listed below:

- unreinforced masonry walls adjacent to safety-related equipment may fall and impact safety-related equipment or cause loss of function of such equipment,
- fire extinguishers may fall and impact or roll into safety-related equipment, or spurious actuation of the fire extinguish system may happen,
- inadequately anchored or braced equipment as vessels, tanks, heat exchangers, cabinets etc. may overturn, slide and impact adjacent safety-related equipment,
- equipment carts, chains, air bottles, welding equipment etc. may roll into, slide, overturn, or otherwise impact safety-related equipment,
- storage cabinets, office cabinets, files, bookcases etc. located, for instance in control rooms, may fall and impact adjacent safety-related equipment,
- break/damage of non-safety related piping, tanks, heating may cause spray, flood and loss of function of the safety related systems,
- flexible piping, cable trays, conduits, and heating, ventilation and air-conditioning (HVAC) ducts may deflect and impact adjacent safety-related equipment,
- anchor movement may cause breaks in nearby piping, cable trays, conduits, HVAC ducts etc. that may fall or deflect and impact adjacent safety-related equipment,
- emergency lights and lower ceiling panels can fall down and damage safety-related equipment free crane hooks may bang safety-related equipment in their vicinity,

The plant walk-down is also required for the assessment of the severe earthquake vulnerabilities and design of accident management and mitigation measures, including the identification of the on-site and off-site logistical obstacles.

3.6. Design of upgrading

Design of the upgrading have to be performed according to the design codes and standards and for the design basis earthquake as defined by current licensing basis.

The seismic upgrading are design modifications requiring proper configuration management and regulatory approvals.

3.7. Role of the peer-reviews

All methods presented above for the re-evaluation and re-qualification of the operating plants require specific knowledge and experience and decisively based on the expert judgement. Consequently, the re-evaluation, re-qualification and upgrading of operating plants have to be peer reviewed to provide an independent overview of its adequacy. The recommendations for the peer review are part of the descriptions of the procedures and also given in the (IAEA, 2003, 2009).

4. Pre-earthquake preparedness and post-earthquake actions

4.1. Operating basis earthquake (OBE) exceedance

Operating basis earthquake level is understood as a limit for the continuation of the safe operation. If the plant is designed for two levels of earthquake, i.e. OBE and SSE, the limit of safe operation should be set equal to the OBE PGA measured at free-field, or to the response acceleration level at an appropriate location of the structure, e.g. at containment basement, calculated for the OBE. If the acceleration is crossing the set level the reactor protection system is actuated automatically. An automatic seismic trip system could be designed in accordance with the concept of the reactor protection system design with regard to the instrumentation, redundancy and the logic of the generation of actuating command. The system design should eliminate as much as possible the spurious trips. There are different concepts for selection of the trigger level. A "high level" trip could be set based on some per-cent of the SSE (usually chosen as greater than 60% of the SSE level) and could be designed to minimize spurious trips due to after-shock and low acceleration earthquakes. A "low level" trip would be set to activate on the compressional waves (P waves) when this first arrival caused displacement or acceleration greater than the calculated maximum allowable P wave for an OBE. The decision on the OBE exceedance per acceleration level crossing could be considered as traditional. Considerations have been made regarding advisability of the automated reactor shutdown in case of small earthquakes (Cummings, 1976, IAEA, 1995). There are plants and sites in low and moderate seismic activity regions where an automatic PGA or acceleration level triggered shutdown can be caused by practically harmless ground motions. There are plants that are practically not designed for two levels of earthquakes just upgraded to comply with the SSE related requirements. For these cases the U.S. NRC Regulatory Guide and 1.12, 1.166 and 1.167, and the IAEA as well as the NRC documents on the "Advisability of seismic scram" provide guidance; see the also the (IAEA, 1995).

At the plant in the moderate seismicity regions the operational limit related to the OBE exceedance is formulated in terms of cumulative absolute velocity and spectral amplitude of the acceleration and velocity response spectra measured at the free field; see (NRC, 1997a, 1997b and 2000; EPRI, 1988; 1989, ANS 2002). According to U.S. NRC Regulatory Guide 1.166 the OBE exceedance criteria are as follows:

"The OBE response spectrum is exceeded if any one of the three components (two horizontal and one vertical) of the 5 percent of critical damping response spectra generated using the free-field ground motion is larger than:

1. The corresponding design response spectral acceleration (OBE spectrum if used in the design, otherwise 1/3 of the safe shutdown earthquake ground motion (SSE) spectrum) or 0.2g, whichever is greater, for frequencies between 2 to 10 Hz, or
2. The corresponding design response spectral velocity (OBE spectrum if used in the design, otherwise 1/3 of the SSE spectrum) or a spectral velocity of 6 inches per second (15.24 centimeters per second), whichever is greater, for frequencies between 1 and 2 Hz."

The CAV is defined for each component of the free-field ground motion as

$$CAV = \int_0^T |a(t)| \, dt, \tag{8}$$

where a(t) is a component of the ground acceleration, T is the duration of the strong motion.

There are certain rules for the numerical calculation of the CAV: (1) the absolute acceleration (g units) time-history is divided into 1-second intervals, (2) each 1-second interval that has at least 1 exceedance of 0.025g is integrated over time, (3) all the integrated values are summed together to arrive at the CAV.

The CAV check is exceeded if any CAV calculation is greater than 0.16 g-seconds.

If the response spectrum check and the CAV check were exceeded, the OBE was exceeded and plant shutdown is required.

4.2. Seismic instrumentation

The seismic instrumentation has two important roles:

- to provide information for the decision on OBE exceedance
- to register the plant response for the post-event evaluation of the plant condition.

The instrumentation providing response records for the evaluation post-event condition of the plant should be installed at most important/significant location of the structures and main components.

The instrumentation for the judgement on the OBE exceedance has to be designed and fitted to the concept of limitation of the operation in case of earthquake. The instrumentation and voting logic for automatic scram should have the same structure, redundancy etc. as the reactor protection system.

For example the Soviet designed SIAZ (System of Industrial Antisesmic Protection) system initiating automatic reactor scram consists of nine tri-axial accelerometers in three independent systems with independent electric power supply and two sets of them. Contrary to this the tri-axial accelerometers for evaluation of OBE exceedance via CAV and response spectrum criteria should be installed at protected free-field locations.

Regarding design and installation of seismic instrumentation see (NRC, 1997c) and (IAEA, 2003b).

4.3. Development of emergency operational procedures

Activities that have to be executed during earthquake should be defined and adequate emergency operational procedures for accident prevention should be developed for the case of earthquake. The documents (NRC, 1997a, 1997b and 2000; EPRI, 1988; 1989, ANS 2002) and the (IAEA, 2011) provide guidelines for the development of the procedure.

The development of earthquake related severe accident management guidelines can be performed on the basis of severe accident oriented studies (Section 3.4) and IAEA documents (IAEA, 2009b).

5. Implementation example – Seismic safety programme at Paks NPP

5.1. Basic principles and outline of the programme

The case of Paks NPP is significantly different from the cases of other nuclear power plants regarding the initial basis and objective of their seismic safety programmes. Ab'ovo, Paks NPP has not been designed and qualified for the earthquake loads. The reason was twofold: the site seismicity was underestimated and the design basis was set to the MSK-64 intesity 5 that was equal to the intensity of the historically credible earthquake plus one intensity ball. In mid eighties the safety deficiency had been recognised and a programme for the definition of the site seismic hazard had been launched, which had been extended to a comprehensive site evaluation programme, including geological, geophysical, seismological and geotechnical investigations as for design basis regarding the scope and the methodology. The probabilistic seismic hazard assessment had been completed in 1995 and the design basis earthquake had been defined on the 10^{-4}/a non-exceedance level. The Hungarian Regulatory Authority had approved the new design basis, and requested to launch a programme for ensuring the compliance with newly defined design basis.

It was already recognised at the very beginning of the seismic safety programme that a consequent and full scope re-design in line with design codes and standards and subsequent upgrading might be impossible at Paks NPP. Therefore, acknowledging the international practice and IAEA recommendations, the Hungarian authorities allowed the use of methodologies for seismic re-evaluation and re-qualification of operating NPPs, less conservative than the design procedures. Admittedly, in early phase of the implementation of the seismic safety programme of Paks NPP, there was a bloodless hope that the issues at Paks NPP could be managed via application of SQUG/GIP, EPRI deterministic seismic margin method, Seismic Evaluation Procedures of the DoE (see Section 3.2 above).

Contrary to the relative alleviations regarding selection of the re-qualification methodologies, the scope of the seismic safety evaluation and upgrades was set by the regulation as for re-design, covering not only the seismic safety classified SSCs (including interacting items), but the whole scope of safety classified SSCs with three times full redundancy with application of the single failure criterion has been accounted instead of considering a success path and a backup only, etc. Also the process requirements were set as for new design, e.g. the heat removal after the design base earthquake shall be ensured unlimited in time, contrary to the 72 hours requirement applicable in usual margin-type assessment.

The real objective has been clearly understood after performing the first Periodic Safety Reviews in 1999, since the compliance with the just issued Nuclear Safety Regulations (Governmental Decree No 108/1997) requested to be achieved and demonstrated. It was recognised that the methodologies mentioned above does not provide the required for Paks NPP result regarding design base reconstitution, on the other hand they can't be directly applied for VVER plants, certain adaptation was needed for accounting the VVER design features.

The qualification of the Paks NPP have been executed as for the newly defined design basis earthquake by applying procedures and criteria for the new design, combined with the methods and techniques developed for seismic re-evaluation of operating nuclear power plants. The seismic safety programme of the Paks NPP is presented below in Figure 1.

The description of the project as given below clearly indicates the similarities and differences between the programmes as understood in (Gürpinar & Godoy, 1998; Campbell at al, 1998) and programme at Paks NPP (Katona, 2001).

The selection and use of methodologies has been graded in accordance with safety relevance of the system, structure or component.

The Hungarian regulation requires performance of probabilistic safety analyses for internal and external events/initiators. Therefore, after implementing the seismic safety upgrading measures, the achieved level of safety has been quantified via seismic PSA, which provides the value of the CDF and also indicated certain week links to be avoided or accounted.

The implementation of the programme was broken into three phases:

Preparatory phase before 1995: The objective was to prepare the programme in a way that it could be executed within reasonable technical and economical limits (example see in Section 5.4). Learning and trial of the methods were going on simultaneously with the site evaluation. The conservatism had to be handled carefully during the seismic safety assessment. The easy to perform fixes had been designed for preliminary conservative seismic input and implemented. The easy-fix project covers 10184 items for 4 units. The volumes of the works are given in Table 2. Total amount of structural steel used for fixes is equal to 445 tons. Safety related batteries for all four units have been replaced and fixed in the frame of the easy-fix project.

Between 1996-1999: Selection of the methodology, evaluation of as-is seismic capacity and identification of the fixes had been performed.

Design and implementation of fixes 1999-2003: The amount of works is illustrated in Figure 2 and 3 and in Table 3.

The programme was broken down into manageable tasks and projects while the uniformity of the requirements and assumptions between these tasks had been ensured by appropriate quality assurance programme and methodological and criteria documents developed for each task.

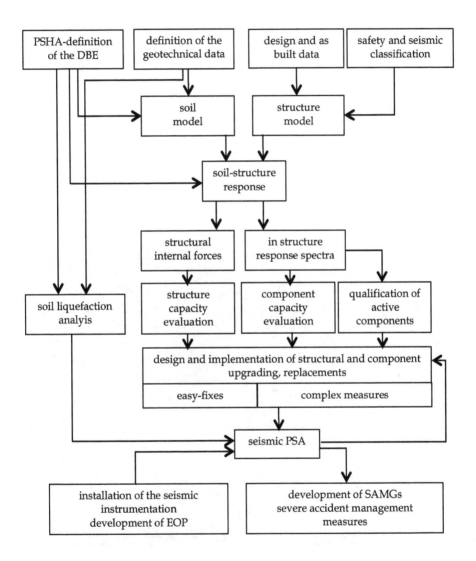

Figure 1. Structure and tasks of seismic safety programme at Paks NPP

Task	number of items	fixes
mechanical equipment	202	anchorages
electrical equipment	465	anchorages
cable trays	2498	anchorages
I&C (cabinets, racks)	2061	anchorages and top bracing
brick walls	281	steel frame fixes
total number of easy-fix items	5507	

Table 2. Tasks and work volumes of the easy-fix project

Figure 2. Viscous-dampers below the steam-generator

Figure 3. Steel-frame bridge between localisation towers for fixing the frames of the reactor hall (columns) in cross-wise direction

Qualification and upgrades	Time frame	Tasks/Volume
Electrical and I&C equipment	1993-2002	Qualifications, replacements
High energy pipelines of primary circuit and equipment	1997-1999	250 fixes (GERB viscous-dampers)
Structure of the turbine and reactor hall	1999-2000	1360 t of steel fixes
Support bridge at localization towers	2000-2001	300 t of steel fixes
Other classified pipelines of primary circuit and the components	1998-2000	760 fixes
Classified piping and components of secondary circuit, fixes of supporting steel structures in the turbine building	2000-2002	160 t of added steel structures 1500 fixes
Other classified pipelines and equipment	2001-2002	80 fixes
Measures identified by seismic PSA	2002-	Fixing the joints in the turbine building, relays qualification

Table 3. Tasks and work volumes of the easy-fix project

5.2. The seismic design basis

5.2.1. Seismic hazard

Full-scope site geological, geophysical, seismological and geotechnical investigation and evaluation has been performed with subsequent probabilistic seismic hazard assessment (PSHA). The methodology is described in (Tóth et al., 2009).

The design base earthquake is defined on the 10^{-4}/a non-exceedance level, taken on the mean hazard curve. Recent Hungarian regulatory requirement is: the design base event has to be defined on the median hazard curve at 0.005 non-exceedance probability for the total lifetime of the plant, which means exactly a 10-4/a frequency for a 50 years operational lifetime (or approximately 10^{-5}/a for a new built).

The horizontal and vertical peak ground accelerations (PGA) are equal to 0.25g and 0.2g respectively. (The PGA correlated to the original design basis seismicity was 0.025g.)

The uniform hazard response spectra were defined for the Pannonian surface (as for a virtual outcrop) below the site. The ground motion response spectra (GMRS) are calculated taking into account the nonlinear features of the soft soil layer covering the site.

The results are shown in Figure 4.

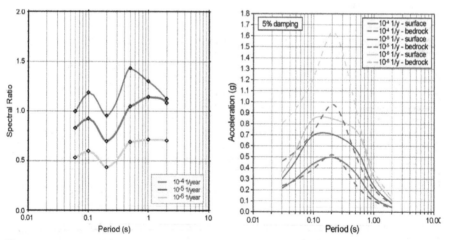

Figure 4. Accounting site mean spectral ratios and mean bedrock and surface UHRS at 5% damping for three different probability levels

The acceptability of the obtained ground motion response spectra for the design base was justified earlier (in 1995) by comparison with deterministically defined 84% response spectra (on the basis of US NRC Draft Guide 1032 issued later as Regulatory Guide 1.165) and recently per U.S. NRC Regulatory Guide 1.208 (NRC, 2005) and ASCE/SEI 43-05 procedure (ASCE, 2005). The latter ensure the avoidance of the cliff-edge effect with respect to the seismic input, since the increase in the amplitudes in ground motion response spectra due to relatively small increase (one order of magnitude) in the exceedance probability is accounted.

The parameters of a 10^{-2}/a non-exceedance level earthquake have also been defined. The PGA is equal to 0.087g in this case. This information is used for certain fatigue type analyses. The response spectrum and cumulative absolute velocity criteria are used for the definition of OBE exceedance; see US NRC Regulatory Guide 1.166 (NRC, 1997a).

5.2.2. Geotechnical conditions

There is Pleistocene layer of 25-30 m covering the site, the upper 12-15 part of which originates from floods and consists of fine structure, well classified sand, while its lower part consist of sandy gravel and gravel. Under the Pleistocene layers, there are various upper Pannonian layers, which are irregularly divided by sandstone ridges. These ridges are cemented to various extents and can be regarded as semi-rock. The 25-30 m saturated young soft soil (~300m/s shear-wave velocity) covering the eroded Pannonian surface at the site is susceptible to liquefaction at the depth 10-15 m. The probabilistic liquefaction analysis performed in 1995 has shown that the best estimate return period of the occurrence of liquefaction exceeds 10000 years therefore the liquefaction was not considered as a design basis phenomenon.

For the seismic PSA purposes, the evaluation of the site effects was extended to very low probabilities ($10^{-4} \div 10^{-6}$/a). According to the seismic PSA the building relative settlement due to the liquefaction is the dominating effect contributing to the CDF just below the design basis probability level. This experience triggered a state-of-the-art analysis of the liquefaction. It was also observed that the uncertainty of the analysis is essential due to uncertainty of soil parameters and the methods.

Liquefaction susceptibility can be expressed in terms of factor of safety FS_{liq} against the occurrence of liquefaction as,

$$FS_{liq} = \frac{CRR}{CSR} \qquad (9)$$

where CRR is the cyclic resistance ration and the CSR is cyclic stress ratio, see Regulatory Guide 1.198 (NRC, 2003). The cyclic stress generated by the given earthquake is as follows (Seed and Idriss 1971):

$$CSR = \frac{\tau_{av}}{\sigma'_{v0}} = 0.65 \cdot \left(\frac{\sigma_{v0}}{\sigma'_{v0}}\right) \cdot \left(\frac{a_{max}}{g}\right) \cdot r_d \qquad (10)$$

where τ_{av} is the equivalent shear stress amplitude, a_{max} is the peak horizontal acceleration at ground surface, g is the acceleration of gravity, σ_{v0} and σ'_{v0} are the total and effective vertical overburden stresses, respectively, and r_d is a nonlinear stress reduction coefficient that varies with depth.

Depending on the method used the value of safety factor varies in rather wide range. The methodologies (Seed & Idriss, 1971, 1982; Tokimatsu & Yoshimi, 1983) used for Paks site resulted in a relative low margin, while the analysis via effective stress method provides much larger margins (Győri et al, 2002).

The building settlement caused by earthquake can affect the underground communications (service water piping and emergency power supply cables) due to relative displacements. This effect will be amplified if liquefaction occurs. The dominant failure mode in the acceleration ranges higher than the design basis is due to the relative building caused by the soil liquefaction. This makes it necessary to re-qualify the underground lines and connections jeopardized by the settlement of the main building or, if it is necessary, to modify them to make their relative movement unimpeded. An advanced probabilistic liquefaction and relative building settlement analysis is going on using an amended soil parameter database in relation to the investigation of beyond design base vulnerabilities performed for severe accident management reasons (Győri et al, 2011).

5.3. Identification of SSCs for safe shutdown – Seismic classification

The procedure for the safe shutdown, cool-down and long-term heat removal of the reactors has been elaborated in two versions (Katona, 2003).

The first version was developed before 1995, when a very conservative guess of the DBE (with PGA 0.35g) was available. For this high demand, a safe shutdown technology was selected that could be realised by the upgrading of the minimum number of systems. It was advisable to select systems for the safe shutdown and cool-down, which are situated in the reinforced concrete containment part of the main building because only this part of the building seemed to sustain the loads. The upgraded and not upgraded systems or parts of systems should be separated in case of earthquake by fast closing valves. The control rods, and boron system would ensure shutdown of the reactor, and the stable subcritical conditions. The reactor should have been cooled-down by the secondary bleed and feed. The long-term heat removal would have been executed through the heat exchangers of the low-pressure emergency core cooling system that should have been modified for the execution of this function. This concept would require modifications in the safety systems and the installation of the great number of valves. Analysing the feasibility issues, it has to be recognised that the implementation of the concept is not only very expensive but it can reduce safety in all other cases than an earthquake because of the modification of the low pressure emergency core cooling system.

Performing the analyses for main building complex, it was recognised that the most critical structure is the gallery building that gives place to several vital systems and I&C equipment. This part of the main building should have been upgraded. Developing the possible technical solution for upgrading the longitudinal gallery building, it turned out that it can be best performed, if the steel frames of the turbine hall and the reactor hall are also fixed. This solution allows the application of structural upgrades that do not require fixes in the over-crowded by equipment and piping gallery building. If it the case, the systems for regular heat removal placed in the turbine hall would be available for heat removal after an earthquake, if their re-qualification is performed. Meantime, in 1995 the site seismic hazard evaluation has been completed which resulted in the DBE with 0.25g PGA. Response and stress calculations made for the newly defined DBE have shown that essential part of the mechanical equipment and pipelines can sustain the DBE demand and the reinforcement of the systems and structures necessary for seismic safety is feasible with reasonable effort.

Theoretical considerations have been made for the evaluation of upgrading effort required for fixing the pipelines and components required for heat removal via systems "as usual", i.e. systems dedicated for emergency cases. It has been assumed that the "as is" seismic capacity of the pipe segments can be treated as a random variable; its value can be expressed by total design capacity multiplied by several factors representing the randomness of the actual design features, floor-response, etc. If it is the case, the calculated "as is" HCLPF values of pipe segments have to be lognormal distributed. If the distribution is known, the parameters of the distribution can be defined on the basis of HCLPF calculations for "as is" conditions and the number of pipe segments requiring fixes can be evaluated. The distributions of "as is" HCLPF values of pipe segments presented in Figure 5 justified the assumptions and made possible the evaluation of upgrading needs.

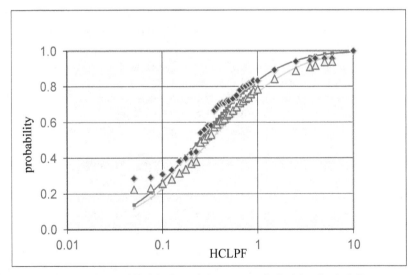

Figure 5. Distribution of "as is" HCLPF values of pipe segments (units 1-4 and units 3-4)

Based on the assessment of fixes of the piping and components, the cost of these fixes turned to be cheaper than the (automatic) isolation of the unreinforced parts of the systems by a great number (more than 100/unit) of fast closing valves.

Consequently, instead of a success path and backup for heat removal, the concept has been chosen that is based on the use of systems devoted for heat removal as per design, taking into account the design philosophy of the VVER-440/213 and widely using the synergy between structural and component fixes.

Meantime, the new nuclear regulation issued in 1997 required the upgrading and qualification of the SSCs enrolled into seismic classes. Moreover, the regulation requested to ensure the adequate capacity of the safety classified SSCs for hazards. Consequently, the scope of the seismic safety evaluation and upgrades was set as for re-design, covering not only the seismic safety classified SSCs (including interacting items), but the whole scope of safety classified SSCs with three times full redundancy, with application of the single failure criterion. The stable and unlimited in time cold shutdown condition have to be ensured after the design base earthquake.

According to the selected concept the sub-criticality is maintained by the shutdown and boron control systems. The cool-down is ensured by secondary-side bleed and feed. The continuous cooling is maintained by the heat removal system. In all redundant trains, the SSCs needed for ensuring these safety functions are fixed and qualified for DBE.

Certain modifications have been implemented for making possible the required functioning, e.g. modification of the venting of the tubes of control assemblies on the reactor pressure vessel head. The systems not required for the safety functions are isolated automatically from the seismically qualified one. The procedure was developed assuming that the plant is in normal operational condition; the outside energy supply (grid) and make-up water source is not available for 72 hours.

Loss of coolant accident is not assumed in consequence of the earthquake; hence, the primary system piping is fixed for DBE according to the design rules. Nevertheless, all redundant safety trains including emergency core cooling systems have been upgraded and qualified for DBE. Consequently the sequences with loss-of-coolant can also be managed, although these are already beyond design base sequences according to the safety philosophy.

On the other hand, the consequences of the small breaks (impulse pipes, drains, air vents) shall be examined from the aspect of the dose limits and containment integrity. The break of small-bore pipes shall be considered in connection with the passive single failures (see article 5.3 of NS-G-2.13). The degree of passive single failures is limited to the break of small-bore pipes (<DN50) and to the leakage of the sealing of pumps or valves.

Those non-safety-classified SSCs have to be also fixed for DBE, failure of which may endanger the integrity or functioning of the safety systems. The possibility of fires and flooding induced by earthquake is also avoided via modification and fixing of the relevant systems, and installing letdown systems for lubricant and Hydrogen.

The systems for the heat removal of the spent fuel and refuelling pools are also fixed and qualified for DBE.

The SSCs have been formally classified: Seismic Class 1 – active systems and components, Seismic Class 2 – passive structures and components (thereafter SCs) needed for ensuring the basic safety functions during and after DBE; Seismic Class 3 – SSCs are those, failure of which may inhibit the safety functions (interacting SCs, falling-on, casing fire or flooding, etc.). Seismic Class 4 – no safety functions and no interaction. Obviously, the scope of seismic safety programme at Paks NPP envelops the scope defined in the international practice for the operating nuclear power plants. Chapter 5 of NS-G-2.13 (IAEA, 2009) only prescribes the re-qualification of the minimal number of SSCs necessary for the implementation of safety functions during and after the earthquake. In case of Paks NPP this concept could not been applied since the design basis had to be reconstructed. Thus the requirements of the IAEA NS-R-1 and NS-G-1.6 were applied. The rational for seismic classification is rather questionable. If the safety related safety classified SSCs have to be designed for the design base external hazards, the basic safety functions would be ensured by these SSCs in case of earthquake too, i.e. no need of the seismic and safety classification.

5.4. Response and strength analysis of structures and components

Two approaches can be accepted for analysis of the soil-structure interaction:

- flexible volume model, flexible volume frequency domain method
- rigid boundary model, rigid boundary time domain method.

In case of the rigid boundary method, the modal damping was limited according to international standards (e.g. KTA 2201.3: 15% for horizontal, 30% for vertical motion). Although the uncertainty of the geotechnical data had been extensively studied, three values of the soil share modulus G_{min}, G_{av} and G_{max}, have to be considered for handling the uncertainty of soil parameters, where $G_{min} = 0.5\ G_{av}$ and $G_{max} = 2.0\ G_{av}$ (according to ASCE-4 (ASCE, 1998), 1.5 G_{av} is acceptable as minimum value).

The analysis of the structural response and capacity of the structures graded approach have been applied, i.e. the modelling and the analysis method have been selected according to the safety relevance of the structure.

The most important building complex is the VVER-440/V213 main reactor building that consists of the reinforced concrete confinement with the localization tower and the attached longitudinal and transversal gallery buildings, as well as the reactor and turbine hall. The most critical parts of the complex structure are the longitudinal and transversal gallery buildings. A method with solution of the equations of motion in frequency domain has been applied for analysis (Katona et al, 1995a).

The secondary buildings are box-shaped structures composed of reinforced concrete prefabricated elements or structures composed of foundation and an upper steel structure. Because of the structural complexity of these buildings, an up-to-date 3D modelling was required. The soil-structure interaction was be modelled by frequency independent soil springs and dampers.

Unique blast tests have been performed for empirical modal analysis of the dynamic behaviour of the main building structures and for the verification of the models developed (Katona et al, 1992, 1993a; Halbritter at al, 1993a). These tests provide good information regarding soil-structure interaction under small-strain excitation.

For optimal modelling of primary system responses, a coupled mechanical and structural model has been developed (Halbritter et al, 1993b; Katona et al, 1993b, 1994, 1999).

The selection of upgrading concept for buildings has been made iteratively. For all options of upgrading, the response and resistance of modified structure has been made and the optimal solution selected via comparison of response and strength achieved. After selection of final upgrading solution the dynamic calculations have been repeated for the modified configuration for justification of the adequacy of the upgrades and development of the floor response spectra. Latter has little importance for the reinforced concrete containment part of the main building complex, but it was essential, e.g. in the gallery buildings. The same iterative procedure has been applied in case of Reactor Coolant System upgrade, and the fixed configuration has been re-calculated for the justification of code compliance of the integrity (Katona et al, 1999).

The methods for evaluation of as-built capacity of structures and components (passive SCs) have been selected in accordance of safety and seismic class, as follows:

-    Safety Class 1 and 2 mechanical components, piping, etc. and Safety Class 2 buildings –
     straightforward design procedure (codes and standards) have been used. For example,
     for the pressure retaining boundaries Class 1 and 2 German design code KTA, and for
     the Class 3 ASME; KTA-ASME comparative study also made, purely elastic approach)
     has been applied;
-    The Class 3 SCs failed to comply with design codes have been generally evaluated
     using realistic assumptions for damping and ductility similar to the (IAEA, 2003).
-    small bore, low-energy pipes – simplified code based method.

The floor response spectra used for the component capacity evaluation was defined according to the design codes (see e.g. ASCE-4-86). However, in case of Class 3 SCs that failed when conservative floor response was used, the calculation was performed for the best estimate floor response spectra (FRS). The best estimate FRS has been obtained either via probabilistic method, or taking into account the inelastic energy absorption, or accounting the equipment-structure mass ratio.

In those cases when the existing supports of pipelines are modified in order to provide adequate seismic capacity, e.g. when the number or type of the supports is changed, it shall be demonstrated on the basis of the relevant nuclear standards (ASME BPVC Section III (ANSI ANS N690) or KTA 3201, 3211, 3205) that the upgraded high energy pipelines and their supports comply with the following criteria:

- The effect of restrained thermal expansion due to the modified pipeline's supports complies with the requirements of the standard;
- The requirements of the standard are met without using the ductility.
- The pipeline's supports comply with the requirements of ASME BPVC Section III NF or ANSI ANS N690 or equivalent nuclear standards (e.g. the German KTA) when they are affected by pipeline reaction due to normal operational conditions (including the restrained thermal expansion) + seismic inertia forces + seismic anchor motion loads related to DBE.

5.5. Qualification of active components

The qualification of active components has been made by several methods:

- Some systems should be replaced or reconstructed for safety upgrading reasons, e.g. the reactor protection system (Siemens Teleperm XS). The new systems and equipment should be qualified and certified by the supplier for the conservatively defined floor response spectra.
- Shaking table testing of sample items.
- Qualification via empirical procedures (GIP, GIP-VVER).

For example, the relays have been qualified by replacing the not to be qualified by new one, shaking table testing of samples for in-rack response spectra (Katona et al, 1995b), experience based method, where it was applicable.

Since the GIP database does not specifically include all the equipment of Paks NPP (manufactured in the Soviet Union or Eastern European countries), it was necessary to apply GIP-VVER (Masopust, 2003) incorporating the knowledge and experience gained during the evaluation of VVER type power plants.

The comparison of 1.5 times bounding spectra (BS) to the floor response spectra is always recommended instead of the comparison of bounding spectra to the ground motion response spectra even below the 12 m level of the building.

5.6. Summary of assumptions, codes and standards and methods

The assumptions accepted for the re-evaluation are summarised in the Table 4. The applicable codes and methods are summarised in the Table 5.

The mixed use of the codes was excluded by careful definition of the evaluation packages. The assumptions, allowable stresses, etc. of the KTA and ASME have been compared.

The operability of active technological components should be qualified by empirical re-qualification procedures or test. The equipment classes and applied empirical qualification methods for active and certain passive components are summarised in the Table 6.

Load combinations	NOL+DBE	
Damping, ductility	Code values or realistic for repeated checking of outliers	
Structural models	Graded approach to the modelling: best estimate if applicable	
Floor response spectra	Conservative design floor response spectra. In specific case best estimate	
Material strength	Minimum values determined by standard	
Capacity evaluation	Design type evaluation	KTA, primary system and vital mechanical equipment and pipelines inside the confinement area
	Margin type evaluation	CDFM+ASME
	Simplified evaluation	Code based simplified procedures
Operability	GIP or GIP-VVER, if applicable, otherwise test	

Table 4. Summary of applicable standards and methods

Equipment	Item	Applicable standards
Passive equipment (tanks, pressure vessels, etc.)	Component body including internal parts	ASME BPVC Section III, Service level D KTA 3201/3211
	Supports	ASME BPVC Section III Subsection NF KTA 3205; Subsection according to Classes.
	Essential nozzles	ASME BPVC Section III, Service level D KTA 3201/3211
	Interactions	GIP, GIP-VVER
Active equipment	Operability	GIP-VVER
	Component body including internal parts	ASME BPVC Section III, Service level D KTA 3201/3211
	Supports	ASME BPVC Section III Subsection NF KTA 3205;
	Essential nozzles	ASME BPVC Section III, Service level D KTA 3201/3211
	Interactions	GIP, GIP-VVER
Pipelines	Pipelines	ASME BPVC Section III, Service level D KTA 3201/3211
	Supports	ASME BPVC Section III Subsection NF KTA 3205;
	Interactions	GIP, GIP-VVER

Table 5. Summary of applicable standards and methods

Equipment classes	Recommended qualification procedure
A. The original twenty classes:	
1. Motor Control Centres	GIP if applicable
2. Low Voltage Switch-gears	GIP-VVER, tests if the item does not fit to
3. Medium Voltage Switch-gears	the database
4. Transformers	GIP-VVER experience data or tests
5. Horizontal Pumps	
6. Vertical Pumps	
7. Fluid-Operated Valves	
8. Motor-Operated Valves	
9. Fans	
10. Air Conditioning Devices	
11. Cooling Devices	
12. Air Compressors	
13. Motor Generators	
14. Distribution Panels	
15. Batteries on Racks	
Equipment classes	Recommended qualification procedure
16. Battery Chargers and Inverters	
17. Engine Generators	
18. Instrument Racks	
19. Sensor Racks	
20. Control Panels and Cabinets	
B. Additional classes:	
21. Relays, Switches, Transmitters, Solenoids, Sensors	Test if the item does not fit to the database
22. Electrical Penetration Assemblies	
C. Additional VVER classes:	
23. Vertical and Horizontal Tanks	Limited analysis, GIP-VVER
24. Vertical and Horizontal Heat Exchangers	
25. Ventilation Ducts	
26. Cable Trays and Conduits	
27. Small and Large Bore Cold Pipes	

Table 6. Summary of qualification methods

5.7. Seismic PSA

The final evaluation of the effectiveness of upgrading measures and justification of the acceptable level of achieved safety in terms of CDF have been made via seismic PSA (Katona & Bareith, 1999; Bareith, 2007; Elter, 2006). The seismic PSA demonstrated that the CDF ensured by the implementation of rather extensive upgrading programme is of order of magnitude 10^{-4}/a. The PSA identified also several week links. For example, the capacity of the joints of the turbine hall structure was found insufficient. Eventual collapse of the turbine building may cause steam and feed-water header ruptures that result in total loss of main and auxiliary feed-water and disables closed loop heat removal through the secondary side. Repeated analysis for the case after implementing the additional measures resulted into CDF value of magnitude of 10^{-5}/a, which is acceptable per Hungarian regulations.

The seismic PSA indicated also that the building settlement of the buildings due to the soil liquefaction jeopardizes the communications (pipes for diesel generator cooling and cables coming from the diesel generators) between the buildings. In the lower acceleration ranges the soil liquefaction that cause settlement of the main building plays dominant role in the occurrence probability of total loss of electric power supply. The studies indicated in Section 5.2.2 are focused on the liquefaction hazard.

The methodology of the seismic PSA applied at Paks NPP complies with the best international practice, see IPEEE NUREG-1407 (NRC, 1991) and (IAEA, 1993). The SPSA was developed on the basis of extensive PSA experience and existing PSA models for Paks NPP and information from newly performed response and strength analyses and qualification effort of the plant and plenty of walk-downs.

5.8. Seismic instrumentation and seismic EOPs

In case of an earthquake, the reactor is shutdown either by reactor protection system due to malfunctions, or manually by the operator based on the criteria of CAV and response spectrum criteria for OBE exceedance. The OBE-exceedance criteria is set CAV=0,16gs and response spectrum in the amplified range less than 0,2g. The seismic instrumentation and the pre-earthquake preparedness and post-earthquake actions are defined via adaptation of the IAEA NS-G-1.6, US NRC Regulatory Guide and 1.12, 1.166 and 1.167 respectively. Selection of the OBE exceedance criteria at Paks NPP was based on the analysis of the frequency of expected events, probability and consequences of spurious signals.

It has to be noted that the implementation of the concept and methodology for OBE-exceedance was not a simple copy-paste; it has been adapted to the conditions of Paks NPP. At Paks NPP, if the g≥g_{set} measured at the base mat, the non-upgraded part of certain systems will be automatically isolated from the upgraded one by quick-closing valves. These system's parts do not have function during and after an earthquake and the separation will not disturb the operation either. In the same time, there is also a signal for control room. If an earthquake happens, there are two possible cases:

- The plant will be shutdown automatically due to disturbances, initiating event(s) caused by the earthquake and the sequence of actuations will depend on the initiating event. Further operator actions depend on the plant condition. The operator actions are defined by EOPs and trained on the simulator.
- If the plant remains in operation after earthquake, the decision on OBE exceedance will be made by operator based on CAV and response spectrum criteria. The plant will continue to operate or will be shutdown if OBE exceeded. If the reactor scram initiated but the OBE hasn't been exceeded, the restart has to be performed after predefined testing and walk-downs.

The seismic recording systems composed from tri-axial accelerometers that are installed at critical locations of structures and main components, provide information for the post-event evaluation of the plant condition.

5.9. Summary of the methods applied at Paks NPP

The comparison of the seismic re-evaluation methods "as usual" and the methods applied at Paks NPP is shown in the Table 7.

	Seismic Margin	Seismic PSA	DB reconstitution at Paks NPP	S PSA at Paks NPP
Input	RLE	Hazard curve	*PSHA: median 10⁻⁴/a non-exceedance, site specific GMRS (UHRS), nonlinear soil, DRS as per Reg. Guide 1.208*	*Hazard curve; UHRS; Nonlinear soil for GM; Analysis of liquefaction;*
Scope	Success + backup path	Event tree/ fault tree	*All Safety related (+ interacting)*	*Event tree/ fault tree modelling*
Structural response	Median structural response, frequency shifting	Probabilistic structural response	*Conservative structural response, (G_{min}, G_{max}, G_{av}), conservative FRS, median FRS in limited cases, Class 3 outside of containment*	*Detailed information available, from the previous works*
Screening	Walk-down and screening per margins criteria, experience-based	Walk-down and screening per fragility estimations	*Walk-down and screening per margins considerations and GIP, GIP-VVER, only the bounding spectrum criterion was accepted.*	*Based on the extensive previous works*
Evaluation, qualification	Analysis of selected SCs (CDFM)	Selected fragility calculations. Median capacities+ log standard deviations	*As per new design for Class 1 and 2 SCs, realistic damping and ductility for Class 3; Testing, GIP, GIP-VVER and replacement for active*	*Detailed information available + fragility development based on the results of the performed analyses; Containment + Liquefaction*
Relay qualification	Screening; qualification of outliers	Screening + limited fragility	*Qualification per screening Test or replacement*	*Screen and limited fragility development*
Modifications	Upgrades if needed	Risk informed upgrades	*Replacements, upgrading per design requirements*	*Certain additional needs for upgrades identified*
Results	Plant level HCLPF	CDF	*Design basis reconstituted*	*Weak links, CDF and its uncertainty evaluated*

Abbreviations used in the Table 7: GMRS, UHRS, DRS – ground motion, uniform hazard and design base response spectra respectively, FRS – floor response spectra; G_{min}, G_{max}, G_{av} – maximum, minimum, and average values of the soil shear modulus; CDFM – Code Deterministic Failure Margin (see Section 3.2.2).

Table 7. The seismic re-evaluation methods as usual and the methods applied at Paks NPP

6. Maintaining the seismic qualification during operation

6.1. Modification design and procurement of equipment

During modifications, replacements and reconstructions, design and procurement can be executed by complying with the seismic safety requirements corresponding to the seismic and safety classification. These processes are part of the configuration management and subject to authority approval. There is a database for seismic and safety SSCs. A procedure exists ensuring the adequacy of the design and procurement specifications.

6.2. Operation and maintenance aspects

Adequate maintenance and status monitoring programs are in place in order to maintain the required status of elements classified from the point of view of seismic safety and requiring maintenance, e.g. the anchorage of piping and components, damping devices. Maintenance of the qualification for earthquake is also part of the ageing management programmes.

6.3. Seismic housekeeping

The proper housekeeping is not irrelevant from the point of view of seismic safety. The following actions have to be required:

- Restoring of fixing elements of cabinets and racks after maintenance
- Restoring of the anchorages, fixing bolts, pipe hangers and the damping devices requiring maintenance and review,
- Appropriate fixing of maintenance devices stored in the plant area.

7. Periodic safety reviews

According to the IAEA Safety Standard NS-G-2.10, the aim of the Periodic Safety Review (PSR) is "to determine by means of a comprehensive assessment of an existing nuclear power plant: the extent to which the plant conforms to current international safety standards and practices; the extent to which the licensing basis remains valid; the adequacy of the arrangements that are in place to maintain plant safety until the next PSR or the end of plant lifetime; and the safety improvements to be implemented to resolve the safety issues that have been identified." Regarding external hazards the objective of the review of hazard analysis is to determine the adequacy of protection of the nuclear power plant against internal and external hazards with account taken of the actual plant design, actual site characteristics, the actual condition of SSCs and their predicted state at the end of the period covered by the PSR, and current analytical methods, safety standards and knowledge.

The period of the PSR is generally ten years. During ten years the knowledge and understanding of the site hazard may develop and a feedback from experiences of other plants may motivate review and upgrading programme. As it can be seen from the experiences of plenty of nuclear power plants, the seismic safety, just like the safety in general, is not a static thing and it covers the whole life cycle of the facility.

Author details

Tamás János Katona
Nuclear Power Plant Paks Ltd., Hungary

8. References

ASME (2008) Standard for Level 1/Large Early Release Frequency Probabilistic Risk Assessment for Nuclear Power Plant Applications, ASME/ANS RA-S–2008,

ANS (2002) Nuclear Plant Response to an Earthquake, Rep. ANSI/ANS-2.23-2002, ANS, La Grange Park, IL (2002)

ASCE (1998) ASCE 4 Seismic Analysis of Safety-Related Nuclear Structures and Commentary (ASCE 4-98) (Standard No. 004-98), American Society of Civil Engineers, 1998

ASCE (2005) Seismic Design Criteria for Structures, Systems, and Components in Nuclear Facilities", ASCE/SEI 43-05, 2005

Bareith, A. (2007) Use of Insights from Seismic PSA for NPP Paks, *Proceedings of the Specialist Meeting on the Seismic Probabilistic Safety Assessment of Nuclear Facilities, 336 p, 14 Nov 2007, p. 66-75*, NEA-CSNI-R--2007-14,

Budnitz, R.J. et al. (1985) An Approach to the Quantification of Seismic Margins in Nuclear Power Plants, Lawrence Livermore National Laboratory, NUREG/CR-4334

Campbell, R. D., et al. (1998) Seismic re-evaluation of nuclear facilities worldwide: overview and status, *Nuclear Engineering and Design* 182 (1998) 17–34

Cummings G.E. et al, (1976) Advisability of Seismic Scram, UCRL-52156, June 30, 1976

DoE (1997), Seismic Evaluation Procedure for Equipment in U.S. Department of Energy Facilities, DOE/EH-0545, March, 1997, Available from: http://www.hss.energy.gov/seismic/

Durga, R. K. et al. (2009) Uncertainty Analysis Based on Probability Bounds (P-Box) Approach in Probabilistic Safety Assessment, *Risk Analysis*, Vol. 29, No. 5, pp. 662-675, 2009, DOI: 10.1111/j.1539-6924.2009.01221

Elter, J. (2006) Insights from the seismic probabilistic safety analysis of Paks Nuclear Power Plant, International Conference on Reliability, Safety and Hazard, Mumbai 2005 (ICRESH05), in *Reliability, Safety and Hazard: Advances in Risk-informed Technology*, Editor: P.V. Varde, 2006, pp. 381 387.

ENSREG (2012), European Nuclear Safety Regulators Group, EU Stress Tests, Available from: http://www.ensreg.eu/EU-Stress-Tests

EPRI (1988) A Methodology for Assessment of Nuclear Power Plant Seismic Margin, Electric Power Research Institute, NP-6041,

EPRI (1989) Guidelines for Nuclear Plant Response to an Earthquake, Rep. EPRI-NP-6695, EPRI, Palo Alto, CA (1989)

European Commission (2011) Technical Summary of the national progress reports on the implementation of comprehensive risk and safety assessments of the EU nuclear power plants, Brussels, 24.11.2011 SEC(2011) 1395

Ewers, J., et al. (1993) Time-versus-Frequency Domain Analysis of Nuclear Power Plant Building Structures, *11th International Conference on Structural Mechanics in Reactor Technology*, August 14-20 1993, Stuttgart, Germany

Gürpinar, A. and Godoy, A. (1998) Seismic safety of nuclear power plants in Eastern Europe. *Nuclear Engineering and Design*, Volume 182, Issue 1, 2 May 1998, Pages 47–58

Györi, E. et al. (2002) Site Effect Estimations with Nonlinear Effective Stress Method at Paks NPP, Hungary. In: *EGS XXVII General Assembly*, Nice, France, 2002.04.21-2002.04.26. Paper 4033

Györi, E. et al. (2011) Earthquake induced subsidence and liquefaction studies for Paks site, *Acta Geod. Geoph. Hung.*, Vol. 46(3), pp. 347–369 DOI: 10.1556/AGeod.46.2011.3.6

Halbritter, A. et al. (1993a) Dynamic Response of VVER-440/213 PAKS Nuclear Power Plant to Seismic Loading Conditions and Verification of Results by Natural Scale Experiments. In: Godoy A, Gürpinar editors, *Proceedings of the SMiRT-12 Conference Seminar No. 16 on Upgrading of Existing NPPs with 440 and 1000 MW VVER type Pressurized Water Reactors for Severe External Loading Conditions*. Vienna, Austria, 1993.08.23-1993.08.25. Vienna: IAEA, pp. 534-568.

Halbritter, A. et al. (1993b) Structural Dynamic Response of the Primary System of the VVER-440/213 PAKS NPP due to Seismic Loading Conditions. In: Godoy A, Gürpinar editors, *Proceedings of the SMiRT-12 Conference Seminar No. 16 on Upgrading of Existing NPPs with 440 and 1000 MW VVER type Pressurized Water Reactors for Severe External Loading Conditions*. Vienna, Austria, 1993.08.23-1993.08.25. Vienna: IAEA, pp. 569-582.

Halbritter, A., et al. (1994) "Requalification of the dynamic behavior of the primary system of the VVER 440/213 at Paks", 9th *European Conference on Earthquake Engineering*, August 27- September 2, 1994 Vienna, Austria

IAEA (1993) IAEA-TECDOC-724, Probabilistic safety assessment for seismic events, IAEA, Vienna, 1993, ISSN 1011-4289

IAEA (1995a) A Common Basis for Judging the Safety of Nuclear Power Plants Built to Earlier Standards, INSAG-8, Vienna, 1995, ISBN 92-0-102395-2

IAEA (1995b) Seismic evaluation of existing nuclear facilities, SMiRT-13 post-conference seminar No. 16, Proceedings of the SMiRT 13 Post Conference Seminar No. 16, Iguazu, Argentina, August 21 - 23, IAEA, Vienna

IAEA (1995c) Consultant Meeting on the Advisability of an Automatic Seismic Scram System in Nuclear Power Plants, 3-5 April 1995, Vienna, Austria

IAEA (2000a) Benchmark Study for the Seismic Analysis and testing of WWER Type NPPs, IAEA TECDOC 1176, IAEA, Vienna, October, 2000, ISSN 1011-4289

IAEA (2000b) Safety Standards Series No NS-R-1, Safety of Nuclear Power Plants: Design, Safety Requirements, IAEA, Vienna, ISBN 92–0–101900–9

IAEA (2002) Safety Standards Series NS-G-3.3, Evaluation of Seismic Hazards for Nuclear Power Plants, IAEA, Vienna, 2002

IAEA (2003a) Safety Reports Series No. 28, Seismic Evaluation of Existing Nuclear Power Plants, ISBN 92–0–101803–7

IAEA (2003b) Safety Standards Series NS-G-1.6, Seismic Design and Qualification for Nuclear Power Plants, IAEA, Vienna, 2003

IAEA (2003c) Safety Standards Series No NS-G-2.10 Periodic safety review of nuclear power plants: Safety Guide, IAEA, Vienna, 2003, ISBN 92-0-108503-6

IAEA (2007) Preliminary Findings and Lessons Learned from the 16 July 2007 earthquake at Kashiwazaki-Kariwa NPP, 2007, IAEA, Vienna, Available from: http://www.iaea.org/newscenter/news/2007/kashiwazaki-kariwa_report.html

IAEA (2009a) Safety Standards Series No NS-G-2.13 Evaluation of seismic safety for existing nuclear installations international atomic energy agency, Vienna, 2009,

IAEA (2009b) Safety Standards Series No NS-G-2.15, Severe Accident Management Programmes For Nuclear Power Plants, Safety Guide, IAEA, Vienna, 2009, ISBN 978–92–0–112908–6

IAEA (2010) Specific Safety Guide, Seismic Hazards in Site Evaluation for Nuclear Installations, Safety Standard Series No. SSG-9, IAEA (2010), ISBN 978–92–0–102910–2

IAEA (2011) Safety Reports Series No 66, Earthquake Preparedness and Response for Nuclear Power Plants, IAEA, Vienna, 2011, ISBN 978–92–0–108810–9

INPO (2011) Special Report on the Nuclear Accident at the Fukushima Daiichi Nuclear Power Station, INPO 11-005, November 2011, Available from: http://www.nei.org/resourcesandstats/documentlibrary/safetyandsecurity/reports/speci al-report-on-the-nuclear-accident-at-the-fukushima-daiichi-nuclear-power-station

Kassawara, R.P. (2008) Seismic Margins Assessment, *IAEA Workshop on The Effects of Large Earthquakes on Nuclear Power Plants*, June 21, 2008, Kashiwazaki City, Japan

Katona, T. (1995) Description of the ASTS at NPP Paks. In: Advisability of an Automatic Seismic Trip System (ASTS) in Nuclear Power Plants: RER/9/035, IAEA, Vienna, Austria, (1995), pp. 64-78.

Katona, T. (2001) Seismic Safety Evaluation and Enhancement at the Paks Nuclear Power Plant. In: *Workshop on the seismic re-evaluation of all nuclear facilities: workshop proceedings*, Ispra, Italy, 2001.03.26-2001.03.27.

Katona, T. (2003) Seismic upgrading of Paks NPP, *International Symposium on Seismic Evaluation of Existing Nuclear Facilities*, IAEA, Vienna, 2003. Paper IAEA-CN-106/51.

Katona, T. (2010) Options for the treatment of uncertainty in seismic probabilistic safety assessment of nuclear power plants, *Pollack Periodica* 5:(1) pp. 121-136. (2010)

Katona, T. (2011) Interpretation of the physical meaning of the cumulative absolute velocity, *Pollack Periodica*, Volume 6, Number 1/April 2011, pp. 9-106

Katona, T. and Bareith, A. (1999), Seismic Safety Evaluation and Enhancement, at The Paks Nuclear Power Plant. In: *Proceedings of the OECD/NEA Workshop on Seismic Risk,*

NEA/CSNI/R(99)28. Tokyo, Japan, 1999.08.10-1999.08.12. Paris: Nuclear Energy Agency, Paper III-3.

Katona, T. et al. (1992) Experimental and Analytical Investigation of PAKS NPP Buildings Structures. In: *Proceedings of the Tenth World Conference on Earthquake Engineering,* Madrid, Spain, 1992.07.19-1992.07.24. Rotterdam: A.A.Balkema, pp. 1609-1618.

Katona, T. et al. (1993a) Dynamic Analysis of VVER-440 Nuclear Power Plant for Seismic Loading Conditions at PAKS. In: Kussmaul K F (editor) *12th International Conference on Structural Mechanics in Reactor Technology (SMiRT-12).* Stuttgart, Germany, 1993.08.15-1993.08.20. Elsevier - North-Holland, pp. 229-234. Paper K08/4.

Katona, T. et al. (1993b) Structural dynamic response of the primary system of the VVER 440/213 Paks NPP due to seismic loading conditions, *SMiRT-11 Conference Seminar No.15, Upgrading of Existing VVER 440 and 900 Type PWR for Severe Loading Conditions,* IAEA, Vienna, 23-25 August 1993

Katona, T. et al. (1994) Requalification of the dynamic behavior of the primary system of the VVER-440/213 at PAKS. In: Duma G (editor) *Proceedings 10th European Conference on Earthquake Engineering,* Vienna, Austria, 1994.08.28-1994.09.02. Rotterdam: Balkema, pp. 2839-2845.(ISBN:90-5410-528-3 (set)

Katona, T. et al. (1995a) Time versus frequency domain calculation of the main building complex of the VVER 440/213 NPP PAKS. In: Riera JD (editor), *Transactions of the 13th international conference on structural mechanics in reactor technology (SMiRT-13).* Porto Alegre, Brazil, 1995.08.13-1995.08.18. Porto Alegre: Universidade Federal do Rio Grande do Sul, pp. 187-192. Paper K032. Division K: Seismic analysis and design, vol. 3

Katona, T. et al. (1995b) Earthquake design of switchgear cabinets of the VVER-440/213 at Paks. In: Riera J D (editor), *Transactions of the 13th international conference on structural mechanics in reactor technology (SMiRT-13).* Porto Alegre, Brazil, 1995.08.13-1995.08.18. Porto Alegre: Universidade Federal do Rio Grande do Sul, pp. 435-440. Paper K073. Division K: Seismic Analysis and Design

Katona, T. et al. (1999) Dynamic Analysis and Seismic Upgradings of the Reactor Cooling Systems of the VVER-440/213 PAKS 1-4. In: *15th International Conference on Structural Mechanics in Reactor Technology (SMiRT 15).* Seoul, South Korea, 1999.08.15-1999.08.20. Paper K11/3.

Kennedy R. P. and Ravindra M. K. (1984) Seismic fragilities for nuclear power plant risk studies, *Nuclear Engineering and Design,* Vol. 79, 1984, pp. 47–68,

KTA (1990) Design of Nuclear Power Plants against Seismic Events; Part 1: Principles, (Auslegung von Kernkraftwerken gegen seismische Einwirkungen; Teil 1: Grundsätze), Kerntechnische Ausschuss, KTA-Geschaeftsstelle c/o Bundesamt für Strahlenschutz (BfS), Willy-Brandt-Strasse 5, 38226 Salzgitter, Germany

Masopust, R. (2003) Seismic Verification Methods for Structures and Equipment of VVER-Type and RBMK-Type NPPs (Summary of Experiences), *Transactions of the 17th International Conference on Structural Mechanics in Reactor Technology (SMiRT 17),* Prague, Czech Republic, August 17 –22, 2003, Paper # K07-3

McGuire, R. K., Silva, W. J. and Kenneally R. (2001) New seismic design spectra for nuclear power plants, *Nuclear Engineering and Design*, 203 (2001), pp. 249-257

NEA (1998) Status Report on Seismic Re-Evaluation. NEA/CSNI/R(98)5. OECD Publications, 2, rue André-Pascal, 75775 Paris Cedex 16, France

Nomoto T. (2009) Report on the Integrity Assessment of Structures, Systems and Components of the KK-NPP by the JANTI Committee, paper presented at *20th International Conference on Structural Mechanics In Reactor Technology*, Dipoli Congress Centre, Espoo (Helsinki), Finland, August 12, 2009

NRC (1956) 10Code of Federal Regulations Part 50, Domestic Licensing of Production and Utilization Facilities, U.S. NRC, Available from: http://www.nrc.gov/reading-rm/doc-collections/cfr/part050/, last update 2012

NRC (1980) Seismic Qualification of Equipment in Operating Nuclear Plants, Unresolved Safety Issue (USI) A-46," U.S. Nuclear Regulatory Commission, Washington, DC, 1980.

NRC (1983) PRA Procedures Guide, NUREG/CR-2300, U.S. NRC 1983

NRC (1991) Procedural and Submittal Guidance for the Individual Plant Examination of External Events (IPEEE) for Severe Accident Vulnerabilities, NUREG-1407, May 1991.

NRC (1997a) Pre-earthquake Planning and Immediate Nuclear Power Plant Operator Post-earthquake Actions, Regulatory Guide 1.166, (1997)

NRC (1997b) Restart of a Nuclear Power Plant Shut Down by a Seismic Event, Regulatory Guide 1.167, (1997)

NRC (1997c) "Nuclear Power Plant Instrumentation for Earthquakes", Regulatory Guide 1.12, Revision 2, U.S. NRC, March 1997

NRC (2000) Resolution of Generic Safety Issues: Item D-1: Advisability of a Seismic Scram (Rev. 1) (NUREG-0933, Main Report with Supplements 1–33), last reviewed, March 2011, Available from: http://www.nrc.gov/reading-rm/doc-collections/nuregs/staff/sr0933/sec2/d01r1.html

NRC (2007) Regulatory Guide 1.208 A Performance-Based Approach to Define the Site-Specific Earthquake Ground Motion, U.S. NRC, March 2007

NRC (2010) Results of Safety/Risk Assessment of Generic Issue 199, "Implications of Updated Probabilistic Seismic Hazard Estimates in Central and Eastern United States on Existing Plants", U.S. NRC, September 2010, Available from: http://pbadupws.nrc.gov/docs/ML1002/ML100270582.html

NRC (2011) Prioritization of Recommended Actions to be Taken in Response to Fukushima Lessons Learned, SECY-11-0137, October 3, 2011

Prassinos, P.G., Ravindra M.K., Savay, J.D. (1986) Recommendations to the Nuclear Regulatory Commission on Trial Guidelines for Seismic Margin Reviews of Nuclear Power Plants, Lawrence Livermore National Laboratory, NUREG/CR-4482

Richner M., et al (2008) Comparison of PEGASOS results with other modern PSHA studies, *OECD/CSNI-Workshop on Recent Findings and Developments in PSHA Methodologies and Applications*, Lyon, April 7-9, 2008, pp. 573–591.

Seed, H.B. and Idriss I.M. (1971) "Simplified Procedure for Evaluating Soil Liquefaction Potential," *Journal of the Soil Mechanics and Foundations Division*, Vol. 97(SM9), pp. 1249-1273, American Society of Civil Engineers, 1971.

Seed, H.B. and Idriss I.M. (1982) Ground Motions and Soil Liquefaction During Earthquakes, Monograph Series, Earthquake Engineering Research Institute, University of California, Berkeley, CA, 1982.

SQUG (1992) Generic Implementation Procedure for Seismic Verification of Nuclear Power Plant Equipment, Rev. 2, SQUG, 1782, 1992

Tokimatsu, K. and Yoshimi Y. (1983) "Empirical Correlation of Soil Liquefaction Based on SPT – Value and Fines Content," *Soils and Foundations*, Vol 15(4), pp. 81-92, Japanese Society of Soil Mechanics and Foundation Engineering, 1983.

Tóth, L., Győri E. and Katona T.J, (2009) Current Hungarian Practice of Seismic Hazard Assessment. In: *OECD NEA Workshop: Recent Findings and Developments in Probabilistic Seismic Hazard Analysis (PSHA) Methodologies and Applications*: Workshop Proceedings, Lyon, France, 2008.04.07-2008.04.09.pp. 313-344. Paper NEA/CSNI/R(2009)1

Probabilistic Assessment of Nuclear Power Plant Protection Against External Explosions

Heinz Peter Berg and Jan Hauschild

Additional information is available at the end of the chapter

1. Introduction

In recent years new threats required safety assessment experts to reconsider the internal and external loads of nuclear installations, in particular nuclear power plants, focusing not only on internal hazards but also on the destructive power of external hazards such as aircraft crash, flooding including tsunamis, severe weather conditions and also explosions and blasts and their combination which can cause significant damage on the plant's operability, being potentially conducive to severe accidents. The cumulated effects of such external loads include the destruction of buildings and access ways, the debris build-up, the loss of electrical power supply as well as the loss of cooling capacity of the reactor core and the fuel pools.

International experience has shown that internal hazards such as fire and external hazards can be safety significant contributors to the risk in case of nuclear power plant operation. This is due to the fact that such hazards have the potential to reduce simultaneously the level of redundancy by damaging redundant systems or their supporting systems or even to loose all redundancies at once.

This has been strongly underlined by the nuclear accidents at the Fukushima-Daiichi nuclear power plants in March 2011 resulting from a very strong earthquake and a consequential tsunami.

A challenging prerequisite for any effective protection against external hazards is to accurately assess them systematically regarding the adequacy of their existing protection equipment against hazards, in particular those built to earlier standards.

Therefore, comprehensive safety assessments have to be performed in advance with most actual site-specific data und current knowledge of new research results. Potential methods to analyse existing nuclear power plants are deterministic, probabilistic or combined methodologies.

In the past, most of the engineering work in designing safety features for nuclear power plants has been performed on a deterministic basis. Moreover, the use of deterministic safety analysis is still current practice to review the current safety level of operating nuclear power plants against external hazards.

As an observation from other areas, the probabilistic approach provides different insights into design and availability of systems and components supplementing the results from deterministic safety analyses. A more comprehensive risk assessment including the modeling and assessment of external hazards is usually recommended in the frame of periodic safety reviews which are performed about every ten years to get a global picture of the safety level of the nuclear power plant under consideration and which include a comparison to current safety standards and good practices.

In particular in case of probabilistic safety analyses, such an assessment can be very detailed and time consuming. Therefore, it is necessary to have appropriate procedures to screen out, e.g., buildings of a nuclear installation where no further analysis is required or to have a graded procedure for the respective hazard taking into account plant- and site-specific conditions.

The assessment of external hazards requires detailed knowledge of natural processes, along with plant and site layout. In contrast with almost all internal hazards, external hazards can simultaneously affect the whole facility, including back up safety systems and non-safety systems alike. In addition, the potential for widespread failures and hindrances to human intervention can occur. For multi-facility sites this makes the situation even more complex and it requires appropriate interface arrangements to deal with the potential domino effects.

In contrast to other external hazards (e.g., earthquakes, winds, or floods), an explosion has the following distinguishing features:

- The intensity of the pressures acting on a targeted building can be several orders of magnitude greater than these other hazards.
- Explosive pressures decay extremely rapidly with distance from the source.
- The duration of the event is very short, measured in thousands of a second, or milliseconds. This differs from earthquakes and wind gusts, which are measured in seconds, or sustained wind or flood situations, which may be measured in hours.

An explosion is defined as a rapid and abrupt energy release, which produces a pressure wave and/or shock wave. A pressure wave has a certain pressure rise time, whereas a shock wave has zero pressure rise time. As a result of the pressure and/or shock wave, an explosion is always audible. Explosions can be classified into a number of types as illustrated in Figure 1.

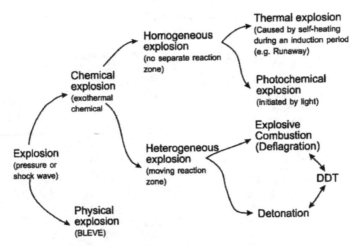

Figure 1. Types of explosion

Explosion is used broadly to mean any chemical reaction between solids, liquids, vapours or gases which may cause a substantial rise in pressure, possibly to impulse loads, fire or heat. An explosion can take the form of a deflagration or a detonation. BLEVE (Boiling Liquid Expanding Vapour Explosion) is a physical explosion also resulting in pressure or shock wave.

The most common type of chemical explosion is the heterogeneous explosion. In heterogeneous explosions, a propagating reactive front clearly separates the non-reacted materials from the reaction products. The reaction front, usually called the reaction zone or flame (front), moves through the explosive mixture as the explosion occurs. In this zone the strongly exothermic reactions occur. Heterogeneous explosions are divided into two types: deflagrations and detonations.

In deflagrations, the reaction zone travels through the explosive mass at subsonic speed, while the propagation mechanism is heat transfer (by conduction, radiation and convection). Reaction zone propagation velocities (flame speeds) of deflagrations may vary over a wide range and so do the corresponding explosion pressures. One example of a deflagration experiment is shown in Figure 2; in this case the deflagration was very short and lasted less than one second.

In some instances, accelerating deflagrations show a deflagration-to-detonation transition (DDT) as shown in Figure 1.

The major characteristic of a detonation is its extremely high speed: the explosion zone moves at a supersonic speed. While, for deflagrations, the flame speeds are relatively low (typically one to several hundreds of metres per second), detonation flame speeds in air can easily reach one to two kilometres per second. The propagation mechanism of a detonation

is an extremely rapid and sharp compression occurring in a shock wave as one can see from
Figure 3.

Figure 2. Experiment of a deflagration according to [1]

Figure 3. Detonation as the strongest type of explosion according to [1]

In contrast to a reversible adiabatic compression, shock compression occurs irreversibly
(non-isotropic), due to the extreme rapidity with which it occurs.

Both types of explosion pressure waves (caused by detonation of liquids or solid explosives
or air-gas mixtures and such pressure waves caused by deflagrations of only air-gas
mixtures) have to be taken into account in the safety assessment of the plant under
consideration.

The first step of the assessment is a screening procedure in order to determine scope and content of the assessment to be performed, the second step is to propose an appropriate approach for those cases where a full scope analysis has to be conducted. In the latter case methods which can be applied to evaluate the probability of occurrence of an external explosion event are, e.g., fault tree analysis, event tree analysis and Monte Carlo simulation.

The presented results show that the probability of occurrence of external explosion pressure waves can be successfully assessed by means of the Monte Carlo simulation.

2. Guidance on assessing external events

Since 2005, a revised guideline for a probabilistic safety assessment [2] as well as revised and extended supporting technical documents [3-4] are issued in Germany which describe the methods and data to be used in performing probabilistic safety assessment in the frame of comprehensive safety reviews.

In these documents, probabilistic considerations of aircraft crash, external flooding, earthquakes and explosion pressure waves are required. Also on international level, new recommendations regarding external hazards including explosions pressure waves and the safety assessments to be performed are recently issued (see, e. g., [5-7]).

For the site evaluation for nuclear installations which will be built in the future safety requirements have been developed [8-9]. In that context activities in the region that involve the handling, processing, transport and storage of chemicals having a potential for explosions or for the production of gas clouds capable of deflagration or detonation shall be identified.

Hazards associated with chemical explosions shall be expressed in terms of overpressure and toxicity (if applicable), with account taken of the effect of distance. A site shall be considered unsuitable if such activities take place in its vicinity and there are no practicable solutions available.

The safety assessment should demonstrate that threats from external hazards are either removed, or minimised or tolerated. This may be done by showing that safety related plant buildings and equipment are designed to meet appropriate performance criteria against the postulated external hazard, and by the provision of safety systems which respond to mitigate the effects of fault sequences.

Explosion pressure waves with relevance to the site can be caused by shipping, fabrication, storage and reloading of explosive materials in closer distances to a nuclear power plant.

These different causes lead to two significant different types of risky situations for the site and the plant which have to be assessed within a probabilistic safety assessment:

1. The explosive material is available as a stationary source in the neighbourhood of the plant under consideration (e.g., a storage facility or a fabrication facility).

2. The explosive material is mobile, i.e. it is shipped in close distance to the plant on the road, by train or on ships along a river or the sea nearby.

In the latter case, the situation is not stable and changes with the varying distances. Moreover, the transport way could be a straight line or a bent which has to be addressed in the calculations - see [10] for a straight road and [11-12] for a bent river.

Usually, a uniformly distributed accident probability is assumed along the transport way. However, in reality the accident probability may increase in junctions or confluences and – in case of rivers and roads – in curves or strictures. Such an example is explained later on in more detail.

Accidents with explosive material are not only theoretical considerations but happen in reality, sometimes with catastrophic consequences.

Data for traffic accidents on rail or road involving explosions are provided in reference [13]. From the total number of accidents (1932) in this database 37% occurred on railways and 63% on roads. The accidents are classified into four different types: release, explosion, fire and gas cloud. The analysis has shown that in the majority of accident gas was released, followed by fires. Explosions appeared in 14% and gas clouds in only 6%. The most frequent initiating event with 73.5% of the accidents result from collisions.

One extremely severe transportation accident took place in June 2009 in Viareggio which resulted in comprehensive safety evaluations [14-15]. Although no industrial plant was damaged in this accident, the potential explosion severity is visible. The accident followed the derailment of a train carrying 14 tank cars of liquefied petroleum gas. The first tank car was punctured after the derailment releasing its entire content that ignited causing an extended and severe flash-fire that set on fire several houses and lead to 31 fatalities.

A more recent accident happened in January 2011 on the river Rhine in Germany, fortunately without any environmental consequences. However, a ship capsized and blocked for many weeks the river for other transportation but, in particular, had the potential to lead to an explosion because – in addition to 2400 tons mainly of sulphuric acid – one tank also contained water and hydrogen.

A further event happened on Mach 11, 2011 on the river Elbe where a transport ship had a damage of an engine and, thus, needed to be anchored outside the regular waterway. One of the questions which arise from this event was if the boundary conditions usually applied and discussed below could be violated because the ship leaves the determined waterway and was, therefore, nearer to the nuclear power plant.

For the respective nuclear power plant comprehensive investigations regarding explosions pressure waves have been performed within the periodic safety review. This includes the identification of the types of ships which are running on the river, the TNT equivalent, the real distance between the ships and the nuclear power plant. Information is in particular based on the information of the Water and Shipping Office Hamburg. This information shows that the biggest tanker ever transported gas on the river Elbe required a maximal safety distance of about 990 m due to the arrangement and size of the tanks and the explosive material loaded according to [18]. This distance is less than the actual distance of 1200 m between the regular waterway and the nuclear power plant and, thus, would not have been led to a hazardous situation for the nuclear power plant, even in the case that the transport ship would have been a gas tanker.

3. Screening process

In a first step, the important areas of the plant are divided into the three classes A, B and C for the analysis of explosion pressure waves to reflect the degree of protection against the impact by the explosion pressure waves. These classes are the same as for the consideration of aircraft crashes [16].

Class A contains systems, where in case of their damages a hazard state directly arises or where an initiating event may occur which cannot be controlled by the emergency cooling system.

Class B contains systems where in case of their damages a hazard state not directly arises, but where an initiating event may occur which is controlled by the emergency cooling system.

Class C contains the safety systems needed for core cooling.

Typical examples of these different classes are [17]:

Examples for class A are the primary coolant circuit, the main steam safety and shut-off valve equipment in case of PWR or pressure relief valves in case of BWR.

Examples for class B are the network connection with the machine transformers and auxiliary power systems (emergency case), the turbine building (main steam line break, loss of the main heat sink, loss of the main feed water) and the switchgear building. The possibility of false signals in the damage control plants with the consequence of a loss of coolant accident has to be considered.

Class C (separated emergency building) consists of buildings that are structurally designed to withstand external influences, including those buildings which are designed against external events. In general a destruction of these systems does not lead to an occurrence of an initiating event. If - in addition to emergency cooling system functions - also further system functions are located in the same building, these assumptions have to be reviewed. If necessary, the results of this review are to be considered in the analysis.

Basic idea in case of explosion pressure waves is a prescribed check if the frequency of core damage states is less than 1E-07 per year for the plant under consideration. This is the case when

- the total occurrence frequency of the event "explosion pressure wave" (i.e. the sum of all contributions from detonation and deflagration) is determined to be less than 1E-05 per year,
- the buildings of classes A and C are designed against the load assumptions shown in Figure 4,

- the safety distances according to the BMI guideline [18] are fulfilled, based on the formula (1):

$$R = 8 \cdot L^{1/3} \tag{1}$$

with

R = safety distance (in m) of the place where the explosive gas is handled from to the respective plant which should be larger than 100 m, and
L = assumed mass of the explosive material (in kg).

It should be noticed that the total mass to be assumed depends on the type of explosive material.

For the case that the prerequisites of this prescribed check are met, no further probabilistic considerations are necessary.

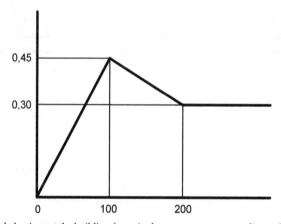

Figure 4. Pressure behaviour at the building for a single pressure wave according to [18]

Otherwise the procedure has to be in accordance with the graded process of evidence regarding explosion pressure waves as presented in Table 1 (see [19]).

Criteria	Extent of analysis
Criterion 1: Occurrence frequency <1E-05 per year Criterion 2: Classes A and C are designed according to load assumptions and safety distances determined in length l_R according to [18]	Verification using the prescribed check
Criterion 1: Not fulfilled Criterion 2: Fulfilled	Conservative estimation of occurrence frequency
Criterion 1: Not fulfilled Criterion 2: Not fulfilled	Detailed probabilistic safety analysis required

Table 1. The graded process of analysing explosion pressure waves

4. Methods as recommended in the German PSA document for nuclear power plants

4.1. Introduction

The German PSA document on methods [3] describes the approaches to be used in the probabilistic safety assessment which have to be performed in the frame of comprehensive safety reviews of nuclear power plants.

One part of this approach is dedicated to the screening process already explained in section 2, the further parts of this document deal in more detail with the occurrence frequency of explosion pressure waves taking into account the site-specific situation, sources of possible explosion pressure waves in the surrounding of the plant, and the procedure for the calculation of occurrence frequencies of accidents during transportation of explosive material by ships, trains or trucks and of accidents of stationary plants near the plant under consideration.

4.2. Assessment

In case that the plant buildings classified as A and C are designed according to the BMI guideline [18] and the safety margins regarding distance and mass of the explosive material are kept, it can be assumed that in the most unfavourable case of an explosion pressure wave event

- no event is initiated which directly leads to a hazard state,
- due to the event explosion pressure wave a system failure occurs in the class B and an initiating event is initiated which can be controlled by the emergency cooling system as designed,
- the emergency cooling system is protected against the effects of the event explosion pressure wave.

In the most unfavourable case, a loss of offsite power with destruction of the secondary plant parts (main heat sink, feed water supply) can be assumed, which occurs with the total occurrence frequency of the event explosion pressure wave. It is assumed for simplifying the analysis that together with the occurrence of this event those systems which are outside of the classes A and C fail.

For the calculation of the frequency of the hazard state, resulting from explosion pressure waves, this initiating event and the incident-controlling functions of the emergency cooling system (stochastic non-availabilities) are to be modelled and quantified in an event tree (or using another appropriate method).

The frequency of the event explosion pressure wave to be chosen is the sum of all contributions of the events detonation and deflagration, as far as they can lead to hazardous states of the plant, resulting from accidents during transportation procedures or the operation of stationary plants in the surrounding of the plant under consideration.

The occurrence frequency of a detonation is several orders of magnitude lower compared with a deflagration [20]. As far as the distance of the area where the deflagration started has a distance larger than 100 m from the plant under consideration (see safety margins in accordance with [18]), no endangerment of the plant buildings has to be assumed.

In case of accidents with materials with the potential of a detonation (in particular explosives, ammunition, gases exothermically disintegrating) the detonation is expected to occur at the accident location, i.e. at a transport route or a fixed industrial installation. Here the approach as provided in formula (2) is applied:

$$H_{E,SMZ} = H_{U,SMZ} \cdot W_Z \tag{2}$$

with

$H_{E,SMZ}$ Annual frequency of a explosion pressure wave by explosives, ammunition or gases exothermically disintegrating in the surroundings of the nuclear power plant,
$H_{U,SMZ}$ Annual frequency of accidents with explosives, ammunition or gases exothermically disintegrating in the surroundings of the nuclear power plant,
W_Z Conditional probability of the ignition in case of an accident.

The deflagration pressure of maximal 10 bar drops over 100 m around a factor 1E04, so that within the power station pressure values within the range of the wind pressures are reached.

In case of explosive gas air mixtures (combustible gases with air; inflammable steams, e.g. also of liquid gas, with air) clouds can appear and a drifting of these clouds from the place where the accident happened into the direction of the plant is possible.

In this situation the deflagration can take place in the area of the plant buildings. The approach applied for this case is described in the following equation [20]:

$$H_{E,GLG} = H_{U,GLG} \cdot W_M \cdot W_D \cdot W_Z \tag{3}$$

with

$H_{E,GLG}$ Annual frequency of an explosion pressure wave by gas air mixtures in the surroundings of the nuclear power plant,
$H_{U,GLG}$ Annual frequency of accidents with combustible gas in the surroundings of the nuclear power plant,
W_M Conditional probability for the development of an explosive gas air mixture in case of an accident with combustible gas,
W_D Conditional probability for drifting of the gas air mixture to the nuclear power plant (as a result of temporal averaging of the arising wind directions),
W_Z Conditional probability of the ignition at the area of the plant.

In a more detailed verification the assumptions introduced can be replaced by plant-specific proofs, considering the different effects of the determined explosion pressure waves.

In the case of a deviation from the BMI guideline [18] partial results of the total occurrence frequency of the event arise which contribute directly to the frequency of the hazard states. These contributions are to be determined by a differentiated view of the assigned explosion pressure waves and their effects.

5. Occurrence frequency of accidents during the transport of explosive materials

One important input for the calculations is the occurrence frequency of accidents during the transport of explosive material with different transportation means. Information has to be gathered from the competent institutions in the respective country. As an example the approach in Germany is shortly described.

5.1. Train accident statistic

According to the accident statistics of the German Railways there was in Germany in the time frame of 10 years no accident of dangerous goods transports with explosive materials. From the zero-error statistics, there is the expectation value for the current admission rate of accidents in Germany in dangerous goods transports by rail with explosive materials ($h_{UEG,B}$):

$$h_{UEG,B} = \frac{1}{2 \cdot 10a} \tag{4}$$

Thus, $H_{UEG,B}$ is defined as

$$H_{UEG,B} = \frac{h_{UEG,B}}{L_{E,B}} \cdot n \cdot l \tag{5}$$

with

$H_{UEG,B}$ yearly frequency of accidents in case of transports of dangerous goods with explosive materials by rail in the vicinity of the nuclear power plant,

$L_{E,B}$ train transport kilometers per year with explosive materials,
n number of transports (trains) per year with an explosive good passing the nuclear power plant,
l section length l along the nuclear power plant (e.g. l = 2 km) which could lead to a hazardous situation for the nuclear power plant.

The section length l can be calculated from

$$l = 2\sqrt{r^2 - a^2} \tag{6}$$

with

a minimum distance of the railway line to the nuclear power plant,

r   radius of the nuclear power plant within which
           a) damages are expected in case of a detonation,
           b) the drifting of a gas-air cloud (deflagration) has to be expected.

5.2. Ship accident statistics

Ship accidents (provided in Germany by the local Waterways and Shipping Directorate) are provided for a defined time period and the river-km and distinguished by the types of accidents. Information with respect to the participation of gas, liquid gas and ammunition shipments to the accident is usually given. The evaluation is performed according to the procedure in [21].

5.3. Occurrence frequency of accidents with explosive materials in stationary installations

In that context, installations such as industrial plants, loading and discharging stations, storage tanks, gas pipelines have to be taken into account according to [3].

In case of natural gas the formation of an explosive gas mixture is only assumed for the accident area because the specific gravity of natural gas is less than the air and drifts of the gas mixture in the direction of the nuclear power plant are therefore excluded.

6. Monte Carlo simulation

6.1. Basics

6.1.1. Monte Carlo simulation

Detailed basics of the MCS like random sampling, estimators, biasing techniques and performance characteristics (e.g. figure of merit / fom) are specified for example in [22] and [23].

In the references [9, 11, 19, 24] the MCS has been applied and verified successfully in order to estimate the probability of external explosion pressure waves.

6.1.2. Estimators in use

As the last event estimator (lee), introduced in [28], is used to predict the probability of an event (e.g. an explosion event), the observed frequency of explosions within the radius r_P is determined. The sample mean probability is

$$\hat{P}_E = \frac{1}{N} \cdot \sum_{i=1}^{N} P_E(i) \tag{7}$$

where $P_E(i) \in \{0, 1\}$ and N = number of trials.

An alternative method is to compute the theoretical probability of an explosion event within the radius r_P in each scenario the wind direction will move the explosive gas mixture to the plant. The advantage over the lee is that each scenario gives a contribution to the probability of occurrence.

By analogy with transport theory, this procedure is called free flight estimator (ffe) also described in [25]. Depending on the accident coordinate (x_i, y_i) and the wind direction ϕ_i in trial i the probability of an explosion event within the radius r_P is given by

$$P_E(x_i, \phi_i) = \exp\left(-\lambda \cdot 1 / v_W \cdot d_1(x_i, \phi_i)\right)$$
$$-\exp\left(-\lambda \cdot 1 / v_W \cdot d_2(x_i, \phi_i)\right) \tag{8}$$

where $d_1(x, \phi)$ and $d_2(x, \phi)$ are the distances between the accident coordinate and the intersection of the wind direction and the plant area with radius r_P.

The intersection coordinates (x_I, y_I) of the wind direction ϕ_i and the plant area with radius r_P are determined by means of

$$x_I^2 + \left(y_i + \tan(\phi_i) \cdot (x_I - x_i)\right)^2 = r_P^2 \tag{9}$$

and

$$y_I = \left(y_i + \tan(\phi_i) \cdot (x_I - x_i)\right)^2 . \tag{10}$$

The sample mean probability is

$$\hat{P}_E = \frac{1}{N} \cdot \sum_{i=1}^{N} P_E(x_i, \phi_i) \tag{11}$$

where N = number of trials.

6.1.3. Biasing techniques in use

If the forced transition method is used (see, e.g., [26]), the next transition is forced to take place within the area (wind direction, distance, time etc.) of interest.

The modified conditional cdf is

$$P(X \le x \mid x_1 < X \le x_2) = F(x \mid x_1, x_2)$$
$$= \frac{F(x) - F(x_1)}{F(x_2) - F(x_1)} . \tag{12}$$

The weight associated to this bias is

$$w^* = F(x_2) - F(x_1) . \tag{13}$$

6.2. Application

The following application is a case study that represents the evaluation of the probability of occurrence of an external explosion pressure wave that takes place near a plant. The probability of occurrence is assessed on the condition that an accident with combustible gas already occurred.

The application is not restricted to a special field of industry; plants of process industry might be in the focus as well as nuclear power plants. The application is depicted in Figure 5. It consists of plant-1 (in the focus of this study), plant-2 (gasholder e.g.), street 1 and 2 (frequented by tank-lorries that carry explosive liquids) and a river (frequented by gas-tanker that carry explosive liquids). The river is subdivided into 6 subsections; each subsection is characterised by an individual length, width and gas-tanker accident frequency.

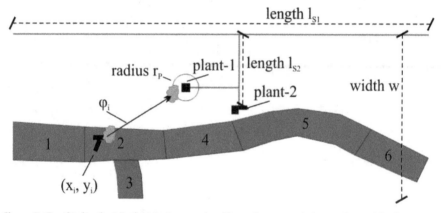

Figure 5. Case study: plant-1, plant-2, river, road and hazardous scenario (gas-tanker accident)

An accident (plant-2, street 1, street 2 or river) at the coordinate (x_i, y_i) may cause the development of explosive gas mixture (gas-tanker accident e.g. - Figure 5).

Depending on the wind direction ϕ_i the cloud of gas mixture can drift to the plant. An ignition of the gas mixture close to plant-1 (within the radius r_P) is in the focus of this study. All relevant application parameters of Figure 5 are given in Table 2.

Description	Parameters
length of street 1: l_{S1}	4,800m
length of street 2: l_{S2}	800m
width w	1,860m
plant 1	area: 10,000m²
radius r_P	150m
plant 2	area: 13,000m²

Table 2. Relevant application parameters

Although the application is described in a generalized way, it incorporates several elements that are typical in order to assess the impact of explosion pressure waves: accident, wind direction, wind speed and ignition.

In the following the example is subdivided into three parts described in sections 6.4 to 6.6: accident at plant-2 (gas holder), accident on street 1 or 2, accident on the river. For each example application the frequency of explosions within the radius rₚ is determined.

The probability of an explosion event within the plant area with radius rₚ is evaluated by means of the Monte Carlo simulation (MCS). In order to make the MCS more efficient biasing techniques are adopted as shown in [26-28]. The algorithm to model and solve the problem is based on the German probabilistic PSA guideline [2] and the supporting technical document on PSA methods [3].

It should be noticed that the events, boundary conditions, parameters and results given in Figures 5 to 14 and Tables 2 to 6 are only example values and do not represent conditions and results of any specific application. However, the described approach is applicable without any general changes by using explicit site and plant specific data.

6.3. Assumptions

The case study depends on the following assumptions:

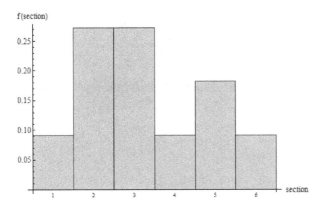

Figure 6. Empirical accident river-section frequencies

- Accident-coordinate:
    - <u>plant-2</u>: Fixed accident-coordinate (x, y) on condition that the accident already occurred.
    - <u>street 1 and 2</u>: Uniformly-distributed accident-coordinate (x_i, y_i) depending on the length l_{s1} and l_{s2} of the streets on condition that the accident already occurred.
    - <u>river</u>: Uniformly-distributed accident-coordinate (x_i, y_i) depending on the subsection of the river on condition that the accident occurred in the river-section i. It is assumed, that the accident frequency is higher in sections with confluences or curves than in straight river-sections.
- The development of explosive gas mixture occurs with fixed probability w_G.
- Empirical-distributed wind direction.
- Empirical-distributed wind speed.
- Exponentially-distributed ignition probability depending on the time.
- An explosion within the radius r_P around the plant is in the focus of this study.

The parameters and distribution models are given in Figures 6 to 8 and Table 3.

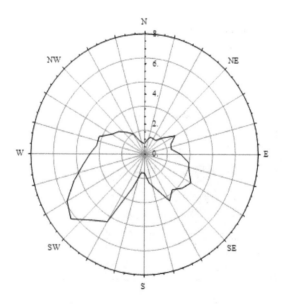

Figure 7. Empirical wind-direction frequencies

Description	Distribution
plant-2: accident (x, y)-coordinate	fixed coordinate
street 1: accident (x, y)-coordinate	U(a, b) depending on length of 4,800m and width of 10m
street 2: accident (x, y)-coordinate	U(a, b) depending on length of 800m and width of 10m
river: accident river-section	empirical
river: accident (x, y)-coordinate	U(a, b) depending on river-section area
development of explosive gas mixture	fixed probability: 0.3
wind direction φ	empirical
wind speed vw	empirical
time τ to ignition	$Exp(\lambda): Exp(0.01\ s^{-1})$

Table 3. Parameters and distribution models

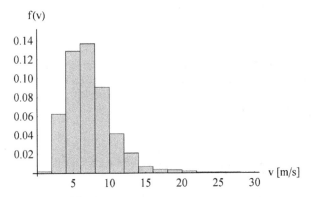

Figure 8. Empirical wind-speed frequencies

6.4. Case study 1 – gas holder accident

The first case study (Figure 9) deals with a gas holder accident at plant-2. The accident at the plant-2 coordinate (x, y) may cause the development of explosive gas mixture. Depending on the wind direction φi the cloud of gas mixture can drift to the plant. An ignition of the gas mixture close to plant-1 (within the radius rp) is in the focus of this study. It is assumed that the accident coordinate (x, y) is fixed. The minimal distance dP2 from plant-2 to plant-1 is approx. 570m. Further relevant application parameters of Figure 9 are given in Table 2 and Table 3.

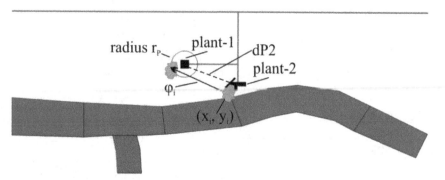

Figure 9. Gas holder accident at plant-2

6.4.1. Analysis

The MCS depends on a sequence of single events:

- Accident (x, y)-coordinate: fixed.
- Development of explosive gas mixture: fixed probability (0.3).
- Wind-direction ϕ: empirical-distributed (Figure 7).
- Wind-speed vw: empirical-distributed (Figure 8).
- Time τ to ignition: Exp(0.01 s^{-1})-distributed.

6.4.2. Results

Different ranges of conditional explosion-probability P_E are depicted in Figure 10. The denotation of the different ranges of the explosion event probability P_E, which is normalised on 1m^2, is as follows: red area (> 1E-07/m^2), orange area ($\leq$ 1E-07/m^2), yellow area ($\leq$ 5E-08/m^2), green area ($\leq$ 1E-08/m^2).

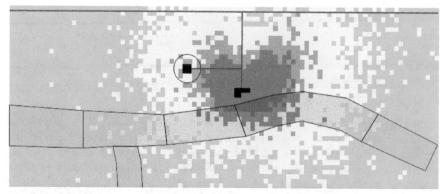

Figure 10. Gas holder accident - ranges of conditional explosion event probability P_E/1m^2

The methods, number of trials, the simulation time and the results like mean value, variance and figure of merit (fom) are listed in Table 4.

The results in Figure 10 reflect the empirical distributed wind-direction, where the cloud of gas mixture is moved in most cases into the direction north-east and north-west.

method	trials	time [s]	mean	variance	fom
MCS-lee	1E05	6.97	3.25E-03	3.24E-03	4.43E06
MCS-lee biased	1E05	25.99	3.26E-03	5.35E-05	7.19E07
MCS-ffe	1E05	7.47	3.28E-03	4.44E-04	3.01E07
MCS-ffe biased	1E05	28.19	3.33E-03	4.49E-05	7.90E07

Table 4. Gas holder accident - conditional probability of an explosion event within the plant area with radius rP

As the different Monte Carlo methods (Table 4) are compared it can be found out that all solutions fit a mean about approx. 3.3E-03 which verifies the results as well as the adopted different Monte Carlo algorithms. If the variance and the figure of merit are regarded the MCS in combination with the ffe and biasing techniques is the most efficient approach.

6.5. Case study 2 – tank-lorry accident

The second case study (Figure 11) deals with a tank-lorry accident on street 1 or street 2. It is assumed that the accident coordinate (x_i, y_i) is uniformly-distributed depending on the length of street 1 and street 2. The minimal distance dS1 from street 1 to plant-1 is approx. 595m and the minimal distance dS2 from street 2 to plant-1 is approx. 605m. Further relevant application parameters of Figure 11 are given in Table 2 and Table 3.

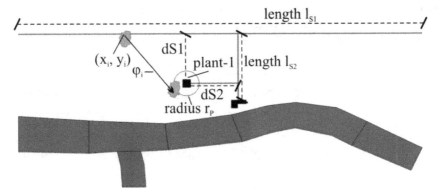

Figure 11. Tank-lorry accident on street 1 or street 2

6.5.1. Analysis

The MCS depends on a sequence of single events:

- Accident (x, y)-coordinate: uniformly-distributed depending on the length and the width of street 1 and street 2 (Table 3).
- Development of explosive gas mixture: fixed probability (0.3).
- Wind-direction ϕ: empirical-distributed (Figure 7).
- Wind-speed vw: empirical-distributed (Figure 8).
- Time τ to ignition: Exp(0.01 s^{-1})-distributed.

6.5.2. Results

Different ranges of conditional explosion-probability P_E are depicted in Figure 12. The denotation of the different ranges of the explosion event probability P_E, which is normalised on 1m^2, is as follows: red area (> 1E-07/m^2), orange area ($\leq$ 1E-07/m^2), yellow area ($\leq$ 5E-08/m^2), green area ($\leq$ 1E-08/m^2). The methods, number of trials, the simulation time and the results like mean value, variance and figure of merit (fom) are listed in Table 5.

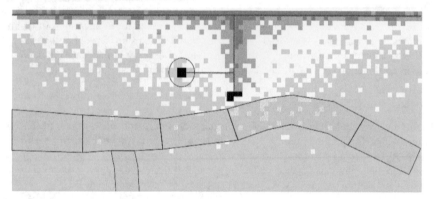

Figure 12. Tank-lorry accident - ranges of conditional explosion event probability P_E/1m^2

method	trials	time [s]	mean	variance	fom
MCS-lee	1E05	6.86	5.40E-04	5.0E-04	2.70E07
MCS-lee biased	1E05	26.69	7.78E-04	6.25E-06	5.99E08
MCS-ffe	1E05	7.04	7.50E-04	8.20E-05	1.73E08
MCS-ffe biased	1E05	27.72	7.70E-04	5.16E-06	6.99E08

Table 5. Tank-lorry accident - conditional probability of an explosion event within the plant area with radius r_P

As the different Monte Carlo methods (Table 5) are compared it can be found out that most solutions fit a mean about approx. 8.0E-04 which verifies the results as well as the adopted different Monte Carlo algorithms. If the variance and the figure of merit are regarded the MCS in combination with the ffe and biasing techniques is the most efficient approach.

6.6. Case study 3 – gas-tanker accident

The third case study (Figure 13) deals with a gas-tanker accident on the river. The river is subdivided into 6 subsections; each subsection is characterised by an individual length, width and gas-tanker accident frequency. It is assumed, that the accident frequency is higher in sections with confluences or curves than in straight river-sections. The accident-coordinate (x_i, y_i) is uniformly distributed depending on the river-section i.

The vertical distances between the plant and the river are between 440m (dR-1) and 780m (dR-2). In the given application ships can reach every location at the river. Further relevant application parameters of Figure 13 are given in Table 2 and Table 3.

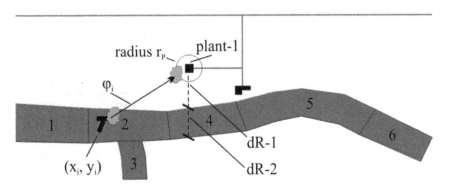

Figure 13. Gas-tanker accident on the river

6.6.1. Analysis

The MCS depends on a sequence of single events:

- Empirical-distributed accident probability depending on the subsection of the river (Figure 6). It is assumed, that the accident frequency is higher in sections with confluences or curves than in straight river-sections.
- Uniformly-distributed accident-coordinate (x_i, y_i) on condition that the accident occurred in the river-section i.
- Development of explosive gas mixture: fixed probability (0.3).
- Wind-direction ϕ: empirical-distributed (Figure 7).
- Wind-speed vw: empirical-distributed (Figure 8).
- Time τ to ignition: Exp(0.01 s^{-1})-distributed.

6.6.2. Results

Different ranges of conditional explosion-probability P_E are depicted in Figure 14. The denotation of the different ranges of the explosion event probability P_E, which is normalised on $1m^2$, is as follows: red area (> $1E-07/m^2$), orange area (≤ $1E-07/m^2$), yellow area (≤ $5E-08/m^2$), green area (≤ $1E-08/m^2$). The methods, number of trials, the simulation time and the results like mean value, variance and figure of merit (fom) are listed in Table 6.

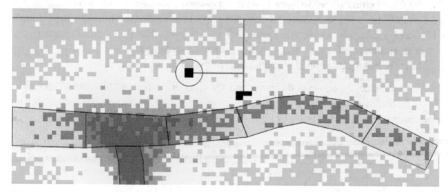

Figure 14. Gas-tanker accident - ranges of conditional explosion event probability $P_E/1m^2$

method	trials	time [s]	mean	variance	fom
MCS-lee	1E05	7.91	1.24E-03	1.24E-03	1.02E07
MCS-lee biased	1E05	28.02	1.30E-03	1.67E-05	2.14E08
MCS-ffe	1E05	8.52	1.31E-03	1.20E-04	9.80E07
MCS-ffe biased	1E05	28.67	1.27E-03	1.35E-05	2.58E08

Table 6. Gas-tanker accident - conditional probability of an explosion event within the plant area with radius rp

Close to the river-sections 2 and 3 the conditional explosion event probability increases, this is due to the higher accident frequency in these sections combined with the specific wind-direction frequencies.

As the different Monte Carlo methods (Table 6) are compared it can be found out that most solutions fit a mean about approx. 1.3E-04 which verifies the results as well as the adopted different Monte Carlo algorithms. If the variance and the figure of merit are regarded the Monte Carlo simulations in combination with the ffe and biasing techniques is the most efficient approach.

6.7. Summary of results

The results of the MCS are evaluated on the condition that the accident already occurred. In order to assess the frequency of occurrence of an external explosion event the frequency of accidents with combustible gas has to be considered. It should be noticed that the results for the frequency of occurrence of an external explosion event will be several magnitudes lower than the results for the conditional explosion event probability given in this paper. Furthermore the events, boundary conditions, parameters and results given in Figures 5 to 14 and Tables 2 to 6 are only example values and do not represent conditions and results of any specific application.

Figures 10, 12, and 14 indicate that the conditional explosion-frequency decreases as the distance to the place of accident increases. This is due to the exponentially distributed ignition probability which depends on the time or the distance to the accident.

The results in Tables 4, 5 and 6 indicate that the conditional probability of occurrence of external explosion pressure waves in consideration of realistic conditions (accident frequency depending on environmental conditions, wind direction & wind speed) can be successfully assessed by means of the MCS.

With the aid of biasing techniques the MCS becomes more efficient, the variance is reduced and the figure of merit (fom) rises. In most cases it can be found out that the solutions fit approx. the same mean which verifies the results as well as the adopted different Monte Carlo algorithms. If the variance and the figure of merit are regarded the MCS in combination with the ffe and biasing techniques is the most efficient approach.

7. Countermeasures to avoid or mitigate the adverse effects of external explosions

Knowledge of the explosion characteristics and the structural impact on buildings of the respective plant is necessary to determine the appropriate countermeasures in order to ensure a safe operation of the nuclear power plant. However, fundamental changes of the plant under consideration are mainly possible only during the design and construction phase.

Basic features of the loads induced on structures by air blasts are described in IAEA Design Guide [5] in terms of a normalized distance that takes into account the amount and type of the explosive charge. The guide presents charts that allow the determination of the peak value of the incident pressure, the total impulse of the positive phase and other relevant design parameters, which are generally used for design or verification purposes of sensitive nuclear structures. For the determination of the resulting actions on structures subjected to a specified blast event, the load-time functions induced by the incident pressure wave must be evaluated in the next step.

In general, it is impossible to protect structures from all man-made and natural hazards. However, assessing the possible damages caused by a defined hazard enables risk-informed decisions about the kinds and number of design changes needed to effectively protect the relevant structures of the nuclear power plants.

This is, in particular, required for the designs of nuclear power plants which are currently under construction or even in the planning phase. Such a hazard assessment has been recently performed [29-30] and for this case a detonation at the highway close to the nuclear power plant has been postulated. For the scenario, the maximum overpressure caused by the explosion has been determined to check if the plant could survive the detonation without damage.

Figure 15 shows the particle velocity field in the pressure wave just before the wave front arrives at the plant under consideration.

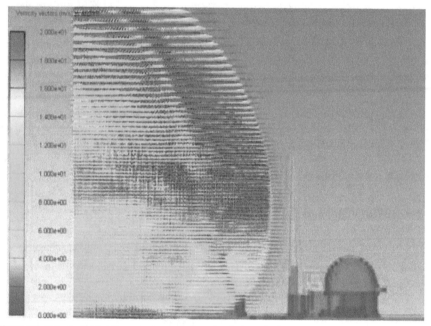

Figure 15. Velocity field at the pressure wave front just before shock wave arrival at the nuclear power plant according to [30]

In case of a plant already operating since several years, the implementation of effective countermeasures is much more difficult or even not possible.

On the one hand, comprehensive calculations can be performed to show that existing assumptions in the calculation provided for the licensing of the plant have been very conservative.

On the other hand, organizational and technical provisions can be taken to reduce the occurrence of an external explosion pressure wave at the plant.

One organizational possibility is to interdict the transport of explosive material, e.g. on a road, in the neighbourhood of the plant. Another solution is to close the road for transit traffic such that the road is only leading to the nuclear power plant.

One technical countermeasure to reduce the explosion frequency on site is the installation of an automatic ignition system placed at a save distance from the site. An assessment has been performed for such an installation which showed that – if the igniters are correctly designed and installed – the shock wave impact after an ignition on the buildings will be limited and will not cause any structural damage.

8. Concluding remarks

The evaluation of external hazards in relation to nuclear power plant design is traditionally considered as a two-step process. The detailed evaluation is preceded by a screening phase where potential scenarios are identified. Many scenarios are screened out on the basis of different criteria, such as distance from the site, probability of occurrence, expected consequence on the plant, or because their effects on the plant are expected to be enveloped by some others. Typically, explosion pressure waves are part of the probabilistic safety assessment as in case of comprehensive periodic safety reviews.

In the German safety guidance document on methods [3] the screening process for the explosion events is explicitly described. The classes of buildings with respect to their protection are the same as for the aircraft crash assessments. Since the updated PSA guideline has been issued in 2005 also requiring the assessment of external events, first practical experience in performing and reviewing the external probabilistic safety assessments are available. One topic is the assessment of the conditional probability of the occurrence of external explosion pressure wave and the discussion of appropriate methods according to the state of the art.

The presented case study and its results (provided in Figures 10, 12, 14 and Tables 4, 5, 6) in the second part of this paper indicate that the conditional probability of occurrence of external explosion pressure waves in consideration of realistic conditions (accident frequency depending on environmental conditions, wind direction and wind speed) can be successfully assessed by means of the Monte Carlo simulation.

As a next step the assessment of explosion events should be extended to include much more realistic boundary conditions regarding

- the extent of the hazard and the explosive gas mixture,
- the ignition probability that depends on environmental conditions [31].

Different ignition models are discussed in [32]. The applied model should be more realistic like the applied exponentially-distributed ignition model; moreover the applicability to integrate the new ignition model into the Monte Carlo algorithm should be given.

Author details

Heinz Peter Berg
Federal Office for Radiation Protection (BfS),Department of Nuclear Safety, Salzgitter, Germany

Jan Hauschild
TÜV NORD SysTec GmbH & Co. KG, Hamburg, Germany

9. References

[1] Shepherd JE (2007) Structural Response to Explosions. 1st European Summer School on Hydrogen Safety, University of Ulster, August 2007.

[2] Federal Office for Radiation Protection (Bundesamt für Strahlenschutz – BfS) (2005) Safety Codes and Guides – Translations: Safety Review for Nuclear Power Plants pursuant to § 19a of the Atomic Energy Act - Guide on Probabilistic Safety Analysis. Bundesamt für Strahlenschutz, Salzgitter, August 2005.

[3] Facharbeitskreis Probabilistische Sicherheitsanalyse für Kernkraftwerke (FAK PSA) (2005) Methoden zur probabilistischen Sicherheitsanalyse für Kernkraftwerke, Stand: August 2005. BfS-SCHR-37/05, Bundesamt für Strahlenschutz, Salzgitter, Oktober 2005 (published in German).

[4] Facharbeitskreis Probabilistische Sicherheitsanalyse für Kernkraftwerke (FAK PSA) (2005) Daten zur probabilistischen Sicherheitsanalyse für Kernkraftwerke, Stand: August 2005. BfS-SCHR-38/05, Bundesamt für Strahlenschutz, Salzgitter, Oktober 2005 (published in German).

[5] International Atomic Energy Agency (IAEA) (2003) External Events Excluding Earthquakes in the Design of Nuclear Power Plant, IAEA Safety Standards Series No. NS-G-1.5, Vienna, November 2003.

[6] International Atomic Energy Agency (IAEA) (2009) Safety Assessment for Facilities and Activities, General Safety Requirements. IAEA Safety Standards Series No. GSR Part 4, Vienna, May 2009.

[7] International Atomic Energy Agency (IAEA) (2010) Development and Application of Level 1 Probabilistic Safety Assessment for Nuclear Power Plants. IAEA Safety Standards Series No. SSG-3, Vienna, April 2010.

[8] International Atomic Energy Agency (IAEA) (2002) External Human Induced Events in Site Evaluation for Nuclear Power Plants, IAEA Safety Standards Series No. NS-G-3.1, IAEA, Vienna, May 2002.

[9] International Atomic Energy Agency (IAEA) (2003) Site evaluation for nuclear installations, Safety Requirements, IAEA Safety Standards Series No. NS-R-3, Vienna, November 2003.

[10] Hauschild J, Andernacht M. (2010). Monte Carlo simulation for modelling & analysing external explosion event probability. Reliability, Risk and Safety – Back to the Future, Proceedings of the ESREL Conference 2010, Rhodes, Balkema, 985 – 989.

[11] Berg HP, Hauschild J (2010). Probabilistic assessment of external explosion pressure waves. Proceedings of the 8th International Probabilistic Workshop, Szczecin, November 2010, 27 – 39.

[12] Hauschild J, Berg HP (2012) How to assess external explosion pressure waves. Reliability: Theory and Applications, Vol. 1, No. 1 (24), March 2012: 50 – 64.

[13] Linkute L (2011). Data on traffic accidents involving fires and explosions. Conference contribution, ISSN 2029-7149 online.

[14] Pontiggia M et al. (2010). Safety of LPG rail transportation in the perspective of the Viareggio accident. Reliability, Risk and Safety – Back to the Future, Proceedings of the ESREL Conference 2010, Rhodes, Balkema, 1872 – 1880.

[15] Manca D, Brambilla S, Totaro R (2010). A quantitative assessment of the Viareggio Railway accident, 20th European Symposium on Computer Aided Process Engineering – ESCAPE20 , 2010, Elsevier B.V.

[16] Berg HP (2005) Screening procedures for the probabilistic analyses of internal and external hazards. Proceedings of the 9th International Conference on Structural Safety and Reliability (ICOSSAR 2005), Rome, 3663 – 3670.

[17] Berg HP (2010) Aircraft crash onto a nuclear power plant. Journal of Polish Safety and Reliability Association, Proceedings of the Summer Safety and Reliability Seminars, June, 21 – 26, 2010, Gdańsk-Sopot, Poland, Volume 1, 27 – 34.

[18] Federal Minister of the Interior (Bundesminister des Innern – BMI) (1976) Richtlinie für den Schutz von Kernkraftwerken gegen Druckwellen aus chemischen Reaktionen durch Auslegung der Kernkraftwerke hinsichtlich ihrer Festigkeit und induzierter Schwingungen sowie durch Sicherheitsabstände. Bundesanzeiger Nr. 179, Sept. 1976 (published in German).

[19] Berg HP, Hauschild J (2011) Assessing external explosions and their probabilities. Journal of Polish Safety and Reliability Association, Proceedings of the Summer Safety and Reliability Seminars, July, 03 – 09, 2011, Gdańsk-Sopot, Poland, Volume 1, 23 – 31.

[20] Federal Minister for Research and Development (1990) Deutsche Risikostudie Kernkraftwerke, Phase B, Fachband 4: Einwirkungen von außen (einschließlich anlageninterner Brände). Köln: Verlag TÜV Rheinland GmbH, (published in German).

[21] Johannsohn G (1971) Gefährdung von Kernkraftwerken durch Luftstoßbeanspruchung von Gasexplosionen und Gasdetonationen als folge von Gastankerkollisionen auf BRD-See- und Binnenwasserstraßen. RWE AG, 1971 (published in German).

[22] Dubi A (2000) Monte Carlo Applications in Systems Engineering. New York: John Wiley & Sons.

[23] Marseguerra M, Zio E (2002) Basics of the Monte Carlo Method with Application to System Reliability. Serving Life-Long Learning, Hagen.

[24] Hauschild J, Andernacht M (2009) Monte Carlo simulation for the estimation of external explosion event probability. Reliability, Risk and Safety, Proceedings of the ESREL Conference 2009, Prague, Balkema, Volume 3, 2139 – 2142.

[25] Marseguerra M et al. (1998) A concept paper on dynamic reliability via Monte Carlo simulation. Mathematics and Computers in Simulation 47: 27 – 39.

[26] Lewis EE, Böhm F (1984) Monte Carlo Simulation of Markov Unreliability Models. Nuclear Engineering and Design 77: 49 – 62.

[27] Hauschild J, Meyna A (2006) Monte Carlo techniques for modelling & analysing the reliability and safety of modern automotive applications. Safety and Reliability for Managing Risk, Proceedings of the ESREL Conference 2006, Estoril, Balkema, Volume 1, 695 – 702.

[28] Hauschild J, Meyna A (2007) Die gewichtete Monte-Carlo-Simulation und ihre Anwendung im Bereich der Zuverlässigkeits- und Sicherheitsanalyse. Düsseldorf: VDI-Verlag, Tagungsband TTZ 2007 (published in German).

[29] Silva de Oliveira R, Cardoso T, Pereira V, da Silva Lima L.G, Gerber B (2011) ANSYS Advantage, Volume V, Issue 3: 34 – 35.

[30] Silva de Oliveira R, Cardoso T, Prates CLM, Riera JD, Iturrioz I, Kosteski LE (2011) Considerations concerning the analysis of NPP structures subjected to blast loading, Transactions of SMiRT 21, 6-11 November, 2011, New Delhi, India, Div-IV: Paper ID# 670.

[31] Hauschild J, Schalau B (2011) Zur probabilistischen Bewertung des EVA-Ereignisses „Explosionsdruckwelle". Tagungsband Probabilistische Sicherheitsanalysen in der Kerntechnik, TÜV SÜD Akademie GmbH, München, 26. – 27. Mai 2011 (published in German).

[32] Drewitz Y, Acikalin A, Schalau B, Schmidt D (2009) Berechnung der Zündwahrscheinlichkeit freigesetzter brennbarer Stoffe im Rahmen einer quantitativen Risikoanalyse. Technische Überwachung, Nr. 9: 35 – 40 (published in German).

Chapter 7

Thermal Reactors with High Reproduction of Fission Materials

Vladimir M. Kotov

Additional information is available at the end of the chapter

1. Introduction

Efficiency of power reactors is determined by the expenses of raw material for its work and by efficiency of heat to mechanical energy transformation.

In general modern nuclear power plants (NPP) use thermal reactors with enriched by isotope ^{235}U uranium (comparing to raw uranium) in the beginning of campaign. This fuel allows having sufficient reactivity margin for obtaining burn-up more than 30-50 MW*day/kg. This reactor type is developed by itself during short period of nuclear power plants development. Raw uranium cannot supply required burn-up even in reactors with best heavy-water moderator. An effective technology of isotope separation was made for military purposes.

Known shortage of thermal reactors is small usage of raw uranium during its work (0.5 – 1.0 %). Stocks of cheap ore for these reactors are enough for 40-50 years at power level of 4000 GW [1].

The next step in nuclear power plants development is suggested usage of fast neutron reactors. This transition is connected with fuel enrichment increase and it supplies possibility of fission reactions on fast neutrons, which produce more secondary neutrons. By solving technical problems at this direction, nuclear power plants supplied with cheap enough fuel for many centuries can be built. But this development direction has one shortage. It is extremely expensive.

When it`s advocates say that high price is because of modern technical solutions shortages, they are half right. Insuperable high price is because of high raw uranium requirement for its start. For using nuclear power plants with total power of 4000 GW, which are supplied for 3000 years, all cheap uranium stocks must be processed in 40-50 years [2]. It is connected with ecological problems and some complexity in non-proliferation of nuclear fission materials.

The best thermal reactors related to high reproduction of fission materials are heavy water moderated reactors. Today there are several types of such reactors. But its potential still is not fully discovered. In CANDU reactors and like-CANDU reactors in the best cases are used fuel on the base of natural uranium as advantage. But achieved burn-up in these reactors is significantly lower than in light water reactors. Besides it, neutron moderation energy and heat leak from channels energy is lost in heavy water. These factors and compactness of light water vessel reactors have caused its leadership in modern nuclear power plants. Now this is shadowing potential performance of heavy water reactors.

There are designs of heavy water reactors, which allow improvement of technical and economical performance of it. In general it is related with use of thorium in fuel.

In fifties gaseous coolant in heavy water reactors have been tested, which allowed to use different values of pressure in reactor and maximal pressure in Rankin cycle. With use of fuel rods, which are much the same design as used in majority of modern reactors, coolant temperature up to 500 °C (EL-4, France) and even a little more (KKN, Germany) was achieved.

Efficiency at temperature, which is similar to achieved, at thermal power plants is more 40%. In the mentioned reactors efficiency is close to 30 % only. Possibly, that this experience served as a reason of transition to high temperature gas cooled reactors with graphite moderator and Briton cycle. Using gas cooled heavy water reactor is not in favor.

If heavy water channels reactors allow better characteristics than existing WWER, PWR, BWR, then it is necessary to know technical solutions, which are needed for this transition.

The purpose of the work is demonstration of thermal reactors development possibility in direction of fission materials reproduction increase, which is sufficient for obtaining burn-up comparable with burn-up of the best modern reactors. This development direction shows that these reactors have high raw uranium usage and can supply high durability of nuclear power plants work at high power with modest requirements in uranium mining. Small amount of fission materials in spent fuel reprocessing is significant advantage. At the same control level it allows less possibility of fission material proliferation. The possibility of reaching the high efficiency coefficient of nuclear plants with the proposed reactors is shown.

2. About usage problems of raw uranium and thorium

In fission reactors can be used uranium and thorium. But only in raw uranium there is fission isotope - ^{235}U. Fission material in thorium is absent, but can be obtained by neutron irradiation (^{233}U). The basic raw uranium isotope - ^{238}U at neutron irradiation becomes fission nuclide - ^{239}Pu.

One of reactor characteristics is raw uranium usage, obtained in its fuel cycle. If core has no conditions for producing ^{239}Pu from ^{238}U, then raw uranium usage is less than isotope ^{235}U portion in it.

In open cycles fuel is once used. A fuel with initial contents of fission and raw isotopes is loaded to a core. At the end of campaign the fission materials contents is decreased and the fuel not used any more. If enriched by ^{235}U uranium is used as an initial fuel, then for raw uranium usage calculations is necessary to calculate raw uranium mass, required for a core loading with enriched uranium [3]. Raw uranium usage I_{sp} is calculated as:

$$I_{sp} = M_{nat} - M_o + M_k;$$ (1)

Where:

M_{nat} – raw uranium mass, required for initial fuel producing;
M_o – mass of initial fuel loading;
M_c – fuel mass at the end of reactor campaign.

Portion of raw uranium usage Q_u, as relative quantity is calculated as:

$$Q_u = I_{sp} / M_{nat};$$ (2)

In closed cycles after the end of campaign fuel is reprocessed for extraction of fission material rests. Different situations are possible.

In the first, the most undesirable situation, not all fission nuclides can be separated from raw nuclides. For example, ^{235}U remains in raw ^{238}U in small amounts so this mix cannot be loaded to a core. Produced ^{239}Pu and ^{241}Pu can be extracted from this spent fuel by chemical methods. Extracted fission materials must be diluted in portion of remained ^{238}U to produce new fuel. Portion of raw uranium usage in this case is:

$$Q_u = \left(I_{sp} + M_{Pu} / C_{dv}\right) / M_{nat}$$ (3)

Where:

M_{Pu} – mass of isotopes ^{239}Pu and ^{241}Pu, extracted from spent fuel and used in new fuel production;
C_{dv} – fission materials contents in initial fuel.

This formula is not taking into account difference between properties of initial and final fission materials and following history of fuel usage. It is estimation. This formula is more precise for condition $M_{Pu} / C_{dv} < 0.5 * M_{nat}$, that characterize modern thermal reactors.

Account of following fuel usage history can be conducted by formula:

$$D_{is} = Y * \left(1 + \psi + \psi^2 + + \psi^{n-1}\right);$$ (4)

Where:

Y – fuel nuclides burn-up during campaign;
ψ – ratio of extracted fuel material mass at the end of campaign to its initial mass;
n – campaign number of this fuel cycle.

The shortage, connected with impossibility of cheap extraction of fission isotope ^{235}U from raw isotope ^{238}U during spent fuel reprocessing, can be overcome by two ways. The first way is in increasing of initial ^{235}U burn-up, so at the end of campaign its contents is negligibly small. Another way is in using fuel with different nucleus charge – using fission isotope of uranium in thorium, fission isotopes of plutonium in raw ^{238}U. Usage of both ways simultaneously is possible.

Because of fission isotopes absence in raw thorium there is no problem with raw thorium usage.

Known stocks of thorium are bigger than known stocks of uranium. It will be ideal if nuclear power industry use both elements in its work. Basic raw material in the present time is uranium. Spending cheap stocks of uranium will lead to necessity of depleted uranium reprocessing from dumps of enrichment plants and spent fuel and thorium usage.

One of the problems with thorium fuel is ^{232}U production, which is source of high energy gamma-ray quanta [4]. One way of this problem solution is usage of automatics at thorium spent fuel reprocessing.

3. Characteristics of hypothetical thermal reactors with high reproduction of fission materials

3.1. Fuel ^{238}U + ^{235}U

High fission nuclides reproduction in thermal reactors is possible if amount of capture acts in fission nuclides is close to amount of absorption acts in raw nuclides.

Dependence of multiplication factor and fission nuclides reproduction coefficient for reactors with different fuel types and no neutron loss is shown on figure 1.

Three types of fuel are shown. Blue lines show data for mix of ^{238}U and ^{239}Pu, red lines for mix of ^{238}U and ^{235}U, brown lines for mix of ^{232}Th and ^{233}U. Continuous lines show multiplication factor, dotted lines – reproduction coefficient. Left lines of each type show data of neatly thermal reactor without epithermal neutron absorption. Right lines show data for reactor with 10% epithermal neutron absorption from total absorptions.

Multiplication factor for such reactor with mix ^{238}U and ^{235}U is calculated as:

$$K = n * \sigma_{5f} * v \ / \ ((1-n) * \sigma_{8a} + n(\sigma_{5f} + \sigma_{5a})); \tag{5}$$

Where:

n – portion of isotope ^{235}U in fuel;
σ_{5f} – fission cross-section of isotope ^{235}U;
σ_{5a} – absorption cross-section of isotope ^{235}U;
σ_{8a} – absorption cross-section of isotope ^{238}U;
v – neutron amount at fission of ^{235}U.

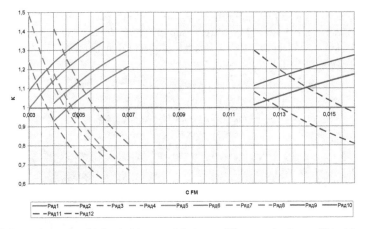

Figure 1. Dependence of multiplication factor and fission nuclides reproduction coefficient for reactors with different fuel types and no neutron loss.

Data for cross-sections and number of secondary neutrons, used in calculations, are taken from [5].

Fission materials reproduction coefficient in initial fuel is calculated:

$$KB = \left(1-n\right) \, {}^* \sigma_{8a} / \; n \, {}^* \sigma_{5f};\qquad(6)$$

From data, which shown at figure 1 for hypothetical reactor, it can be seen that there is diapason of uranium enrichment (from 0.46 up to 0.66 %) in which reproduction coefficient and multiplication factor are more than unity simultaneously. For each of shown variants at reproduction coefficient equal unity multiplication factor is close to 1.1.

To estimate possibility of real reactor work at this diapason of enrichment is necessary to take into account two factors – presence of additional neutron losses in construction materials and neutron leakage, and production influence of secondary fission materials, actinides and fission products, which are sufficient neutron absorbers.

Estimation of these factors in point model of reactor is possible by introduction of neutron loss in construction materials and leakage term in system of equations, describing accumulation and neutron absorption in initial fuel nuclides and additional nuclides produced during reactor work.

Sufficiently precise estimation of reactor campaign characteristics with different fuel types can be made taking into account following nuclides:

- Raw nuclides in fuel – ^{235}U, ^{238}U, ^{232}Th;
- Secondary fission nuclides – ^{233}U, ^{235}U, ^{239}Pu, ^{241}Pu;
- Nuclides of fuel chains – ^{233}Pa, ^{234}U, ^{236}U, ^{237}U, ^{239}Np, ^{240}Pu, ^{242}Pu;
- Actinides – ^{241}Am, ^{242}Am;

- Fission products with high absorption cross-section and atomic mass:
- 99 (Nb, Mo, Tc); 103 (Ru); 131 (Sb, Te, J, Xe); 133 (Te, J, Xe, Cs); 135 (J, Xe); 143 (La, Ce, Pr, Nd); 145 (Ce, Pr, Nd); 149 (Nd, Pm, Sm); 151 (Nd, Pm, Sm, Eu); 152 (Sm); 153 (Sm, Eu); 155 (Sm, Eu, Cd); 157 (Eu, Cd).

In general, for each of nuclides the equation is solved:

$$\frac{dNz}{dt} = \lambda_{z-1} \cdot N_{z-1} - \lambda_z \cdot N_z - \sigma_z \cdot N_z \cdot \Phi; \tag{7}$$

where:

Nz – number of nuclei with charge z in fuel;
λ – decay constant;
σ – neutron cross-section absorption;
Φ – neutron flux in fuel

Calculations of campaign of point reactor model with these conditions are made with the program [6]. The program takes into account fission possibility for ^{238}U, ^{232}Th, ^{233}Pa, ^{234}U, ^{236}U, ^{237}U, ^{239}Np, ^{240}Pu, ^{242}Pu [5].

Results of reactor campaign calculation with 0.47 % initial contents of ^{235}U in fuel and absence of epithermal neutrons absorption in ^{238}U are shown at figure 2. Here (and everywhere in analogous cases) contents of fission nuclides (^{235}U и ^{239}Pu) is normalized on initial contents of ^{235}U in fuel, and fuel power – on its initial value.

For representative comparison of campaigns is made Table 1. It includes:

C FM - the content of the base fission materials at the campaign beginning, %;
F M – Fission materials in fuel;
R M – raw fuel nuclides;
Mode – campaign conducting features:
    Line – the simplest flow of the campaign (without fuel replacements);
    Sup Poz – campaign with the joint work of fuel with different burn-up (look at i.5);
    Sup Poz +^{233}U Gen – campaign with generation of ^{233}U (look at i. 6);
AKM – the absorption of neutrons in construction materials and leakage, %;
Φ – neutron flux, $sm^{-2}s^{-1}$;
T – campaign duration, hours;
Y – fuel burn-up, %.
Rmin – minimum operational reactivity during the campaign, %;
R SZ – integral reactivity at the end of campaign, %;
U3 – portion of ^{233}U fissions from total fissions;
U53 – portion of ^{235}U fissions, which formed from ^{233}U, from total fissions;
U55 – portion of ^{235}U fissions, which is from natural uranium, from total fissions;
Pu9 – portion of ^{239}Pu fissions from total fissions;
Pu1 – portion of ^{241}Pu fissions from total fissions;
Qu op –portion of raw uranium usage in open fuel cycle;
Qu sh – portion of raw uranium usage in closed fuel cycle;

№	C FM	F M	R M	Mode	AKM	Φ	T	Y	R_{min}	R SZ	U3	U53	U55	Pu9	Pu1	Qu op	Qu sh
1	0,47	U5	U8	Line	0.0	9+13	13	1.42	0.002	0.054			29.5	63.6	5.43	2.15	no
2	0,37	Pu9					13	1.63	0.008	0.021				76.6	23.40		
3	1,65	U3	Th			9+13	39	8.78	0.019	0.028	90.9	9.10					
4	0,35	Pu 9, 1	U8				13	1.56	0.002	0.03				73.5	26.50		100
5	1,65	U3, 5	Th		0.0		16	4.09	0.001	0.02						-	
6						5+13	38	4.89	0.028	0.046	90.4	9.60					
7	0,68	U + Pu	U8+Th			9+13	12	1.80	0.009	0.034	37.26	4.08		44.26	14.40		
8						5+13	25	2.15	0.008	0.030	40.54	3.98		42.33	13.14		
9	0,47	U5					25	2.52	0.013	0.018			17.52	68.37	14.10	3.81	-
10	0,712						34	3.92	0.010	0.027			17.07	69.05	13.87	3.92	
11	0,35	Pu 9, 1		Super-position	1.7		15	1.79	0.0003	0.009				73.4	26.60	-	100
12	0,712						29	3.38	0.006	0.024			19.78	67.31	12.90	3.38	
13					2.8		24	2.84	0.008	0.029			23.52	64.91	11.55	2.84	
14		U5	U8		5.2		19	2.41	0.002	0.026			27.72	62.22	10.05	2.41	-
15				Super-Positio	1.7	9+13	44	5.27	0.001	0.017	5.05	0.16	12.70	67.69	14.40	5.27	
16	0,47			n	0.0		34	3.47	0.012	0.027	5.27	0.13	12.70	64.30	17.60	5.26	
17	0,35	Pu 9, 1		+ U3 gene-ration			27	3.13	0.001	0.012	4.10	0.10		72.83	22.99		100
18	,17+,35						29	3.51	0.001	0.017	1.89	0.04	4.69	71.22	22.16		14.28
19	,35+,35				1.7		34	4.22	0.005	0.024	2.12	0.06	7.78	69.14	20.90		8.58
20	,2+,41	U5+Pu 9, 1	U8(1.2)				24	3.49	0.004	0.014	2.28	0.04	5.52	69.62	22.54	-	12.12
21	,31+,39		U8(1.1)	Super-position			24	3.34	0.005	0.025			8.35	69.24	22.41		7.61
22	,35+,35		U8				24	3.08	0.016	0.037			10.69	67.46	21.84		6.27
23	0,712	U5	U8	Line	5.2	9+13	10	1.26	0.001	0.085			45.13	50.91	3.95	1.26	-
24							34	4.12	0.001	0.01	4.79	0.13	16.23	65.66	13.19	4.12	
25	1.65	U3, 5	Th	S P G	5.0	6+13	24	4.35	0.015	0.044	92.58	7.42				-	100

Table 1. Campaign characteristics

Reactor power with constant neutron flux increases during campaign. Power has peak of 28 % and at the end of campaign is 20 % greater than initial value. It is caused by ^{239}Pu accumulation, which has bigger fission cross-section than ^{235}U, and ^{239}Pu contents stabilization with decreasing portion of ^{235}U.

Contents of ^{239}Pu becomes stable during campaign but its value is less than initial contents of ^{235}U in fuel. It shows necessity of taking into account characteristics difference between fission nuclides at calculation of reproduction coefficient. With use of formula (2) and ^{235}U fission characteristics reproduction coefficient equal to unity is calculated. At the same conditions reproduction coefficient for ^{239}Pu is less unity (0.786).

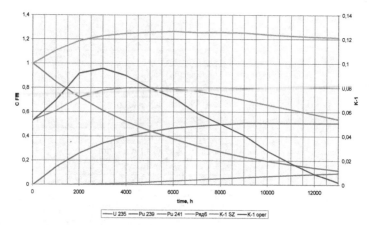

Figure 2. Reactor campaign characteristics with initial contents ^{235}U in fuel 0.47 % and absence of absorption in ^{238}U on epithermal neutrons. (string 1 of Attachment Table 1)

3.2. Fuel 238U + 239Pu

Region with reproduction coefficient equal unity for fuel on the base of mix ^{238}U + ^{239}Pu relocates to less value of fission materials contents comparing to fuel on the base of ^{238}U and ^{235}U.

Multiplication factors for RC=1 is increased.

Reactor campaign characteristics with initial fuel containing ^{238}U and ^{239}Pu are shown at figure 3.

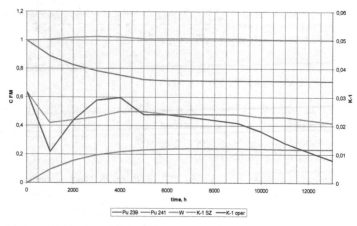

Figure 3. Reactor campaign characteristics with initial contents of ^{239}Pu in uranium-plutonium fuel 0.37 % and absence of absorption in ^{238}U of epithermal neutrons (string 2 of application's table 1).

Portion of ^{239}Pu is 0.37 % from total mass of these nuclides and in ^{238}U there is no absorptions in epithermal region.

Contents of ^{239}Pu during campaign is stable enough, but not equal to initial one and decreasing for ~25 %. Stabilization of ^{239}Pu is reached earlier than in the previous campaign. The role of ^{241}Pu increases. Its amount increases to ~ 1/3 of ^{239}Pu at stationary level. Value of reactivity margin during this campaign is less, but its fluctuation is also less. Despite less contents of fission material in initial fuel slightly higher burn-up is reached after the same work time.

Decreasing of ^{239}Pu amount and reactivity margin in the first hours of campaign is caused by delay of transformation of ^{239}U into ^{239}Pu, and significant neutron losses in the chain of ^{241}Pu. It is important to make comparison with fuel characteristics on the base of mix ^{233}U + ^{232}Th.

3.3. Fuel ^{232}Th + ^{233}U

For the fuel on the base of mix ^{232}Th + ^{233}U the region with reproduction coefficient equal to unity relocates to higher contents of fission materials. Multiplication factor in this region also shows increase comparing to variants with fuel on the base of ^{238}U and ^{235}U.

Reactor campaign characteristics with ^{233}U + ^{232}Th in initial fuel is shown at figure 4. Contents change of ^{233}U during campaign is not big. Reactor power change is also not big. But power is decreasing at the campaign beginning and after that returns to its initial value. Power decrease at the campaign beginning is caused by ^{233}U contents decrease, and return is caused by ^{235}U accumulation. Comparatively small accumulation of ^{235}U is well explained by small neutron absorption cross-section of ^{233}U, from which produces ^{235}U.

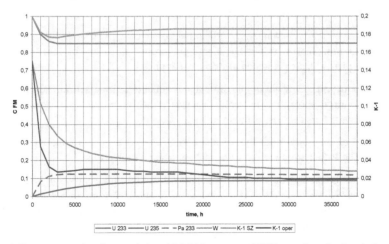

Figure 4. Reactor campaign characteristics with initial contents of ^{233}U in uranium-thorium fuel equal to 1.55 % (string 3 of application's table 1).

Positive reactivity margin in this campaign is decreasing at its beginning because of ^{233}Pa contents increase and its comparatively long half-life. After 2000 work hours reactivity fluctuation is small because all fuel nuclides has stabilized. The reached duration of this campaign is considerably higher than it of the previous campaigns.

4. Analysis of hypothetic reactors campaign

4.1. Variants of campaigns for figures 2, 3, 4

Let us compare some campaign characteristics of three shown variants of reactors with different fuel type, which are significant for its working out for applying in practice. This comparison is needed because the choice of initial conditions for campaign characteristics calculation is, in principle, not equivalent. Comparisons can delete this shortage.

Portion of raw uranium usage is usage rationality of this fuel kind (at this campaign). It is important because uranium is unique natural material containing fission isotope ^{235}U.

Only the first of three considered campaigns can be used in the open cycle, i.e. without nuclides, which were gotten from the spent fuel.

Closed fuel cycles with raw thorium or raw uranium can work without ^{235}U usage. Thus, the complete use of natural uranium, which used at the initial stages of these campaigns, is reached.

At the base of ^{238}U+^{235}U fuel simplest campaign cannot reach high raw uranium usage. Reproduction of fission materials is too small, neutron losses in reactor control elements is too big. Reached burn-up is minimal among all suggested variants.

4.2. Variants with equilibrium fuel

Characteristics of reactors campaign with initial contents of ^{239}Pu and ^{241}Pu sum in the uranium-plutonium fuel, which is equal to equilibrium, which can be formed from the campaign by picture 3, are presented at string 4 of application's table 1.

Fission reproduction coefficient in this campaign is equal to unity practically always during campaign. Neutron absorption in control rods increases up to 3.3 %. Reactor power is practically constant. Its decrease is not more than 1 %.

Optimization of campaign with fuel on the base ^{233}U + ^{232}Th by the same means is also interesting. These data are shown at string 5 of application's table 1.

Equilibrium of fission materials in the campaign with equilibrium contents of ^{233}U, ^{235}U and thorium is achieved after 2000 work hours. Neutron absorption in control rods is 5.8 %. Reactor power decreases to 93.2 % from initial value at 2000 work hours and remains stable. Power decrease is caused by ^{233}U formation and role of ^{233}Pa with long half-time.

Large reactivity margin with increase possibility of neutron losses in construction materials and for leakage is significant difference between this and previous campaign.

4.3. Change of resonant absorption in raw materials

As it can be seen from charts of figures 1 equal reproduction coefficient is achievable with different fission materials contents in fuel. It is done by changing of resonant absorption in raw nuclides. Described results are made for cases with no resonant absorption. Theoretically these characteristics remain constant at the same durability and neutron flux with increase of resonant absorption and achieving the same reproduction coefficient:

- portion of absorption in control rods;
- final contents of fission products;
- final value of reactivity.

Burn-up of fuel nuclides is changed. Reactor work with high burn-up is desirable. For campaign search with increased resonant absorption in raw nuclides data from figures 1 is not sufficient, because it is based on two nuclide campaign when its real number is six in campaign with uranium-plutonium fuel with limitation of ^{240}Pu.

Described cases are not common in reactors with low reproduction coefficient, where resonant absorption in raw nuclides leads to multiplication factor decrease in the campaign beginning and decreases campaign durability.

5. About possibility of practical campaign realization with high reproduction

The basic difference of real reactors is presence of neutron absorption in construction materials and neutron leakage from reactor. These factors can be researched in the described models without reference to the reactor design by additional term insertion for these neutron losses.

Neutron flux values used in previous calculations are not always applicable. Possibility of neutron flux change and its influence on campaign characteristics must be examined.

Arrangements for reactor campaign improvements and its effects should be also considered. Without reference to a reactor design following arrangements can be considered:

- combined use of several fuel kinds (from described above);
- fuel dynamic loading use, which suggests using of several equal fuel portion, each of them works in reactor during specified time, after that it is replaced with next fuel portion. In the cycle each portion is used once;
- zone superposition regime of campaigns with different start moments;
- insertion of excess of fission materials to initial fuel;
- fission materials production with use of control rods with raw nuclides addition (^{232}Th and ^{238}U).

5.1. Neutron flux influence on campaign characteristics

Campaign characteristics calculations, which were presented above, use ~$9*10^{13}$ sm^{-2}sec^{-1} neutron flux. Flux increase allows improving economical issues – decrease of fuel requirement at specified power output. It is especially valuable for the case when during campaign ratio of neutron flux to power is almost constant.

Campaign characteristics calculations for above mentioned fuel types with neutron flux in diapason from $2.5*10^{13}$ to $2.0*10^{14}$ sm^{-2}s^{-1} and campaign durability from 5000 to 40000 hour were carried out. At flux from $2.5*10^{13}$ to 10^{14} sm^{-2}s^{-1} change in final mass of fission materials, control rods absorption and reactivity at the end of campaign are slight for all fuel kinds. Change of its parameters for fuel with raw ^{238}U at neutron flux $2*10^{14}$ sm^{-2}s^{-1} is slight.

Final reactivity for thorium containing fuel under the neutron fluxes more than 10^{14} sm^{-2}sec^{-1} is fast becoming less than zero [7]. Flux rising influence is could be watched at string 5 and 6 of application's table 1, where the one fuel type is used, but it has the different neutrons fluxes in $9*10^{13}$ sm^{-2}sec^{-1} and $5*10^{13}$ sm^{-2}sec^{-1}. At larger flux reactivity is less than zero after 16500 hours, and at smaller flux it is near the 0.03 after 39000 hours. By estimation it becomes zero after 50000 hours of work reactor.

These effects are explained by neutron absorption in ^{233}Pa, which has comparatively high half-time period (27.4 days) and large absorption cross-section (66 barn). However, not everything is so simple. The reactivity in campaign with non-equilibrium fuel (figure 4) and $9*10^{13}$ sm^{-2}sec^{-1} flux remains high during 39000 hours.

5.2. Combined work of several fuel types

Reactor campaign characteristics with 0.64 % initial contents of mix ^{233}U, ^{235}U, ^{239}Pu and ^{241}Pu in uranium-thorium fuel with ^{238}U – 75 % and $9*10^{13}$ sm^{-2}s^{-1} neutron flux are shown at string 7 of application's table 1.

Contents of fission materials is corresponding to its equilibrium contents. Reproduction of fission nuclides is close to unity. Portions of energy release from fission components are:

$$^{233}\text{U} - 39.76\ \%;^{235}\text{U} - 4.26\ \%;^{239}\text{Pu} - 42.21\ \%;^{241}\text{Pu} - 13.76\ \%.$$

That means in fission components from raw ^{238}U occurs ~56 % fissions, and in fission components from raw ^{232}Th ~44 % fissions.

Positive reactivity at the campaign end in this variant is less than with equilibrium fuel on the base of ^{232}Th and ^{233}U. Campaign with the same fuel, but $5*10^{13}$ sm^{-2}s^{-1} flux is presented at string 8 of application's table 1. Reached burn-ups in the both cases are close to each other and it is says about decreasing of negative role of thorium nuclides component fuel.

5.3. Dynamic loading regime use

Most part of absorptions in fission products is in ^{135}Xe. Absorption in ^{135}Xe in high neutron flux (more than $5*10^{13}$ sm^{-2}s^{-1}) is close to portion of ^{135}I formation from fission products. Portion of ^{135}I formation is ~6 %, and total absorption in all fission products is slightly above 10 %.

Peculiarity of ^{135}Xe is its formation from ^{135}I with half-life 6 hours and decay with 9 hours half-life.

In dynamic loading regime [7] fuel works in reactor during time close to ^{135}I half-life, formation of ^{135}Xe and its neutron absorption is minimal. During fuel exposition out of core ^{135}Xe is decaying.

Dependence of neutron absorption portion in ^{135}Xe, which formed during fuel work on work duration and neutron flux and portion of remaining ^{135}Xe after fuel exposition out of core is shown on figure 5.

It can be seen that even in maximal neutron flux (10^{14} sm^{-2}s^{-1}) neutron absorption in ^{135}Xe is 30 % at work time about 8-10 hours.

We can note that portion of remaining ^{135}Xe after fuel exposition practically does not depend on loading characteristics of fuel in reactor.

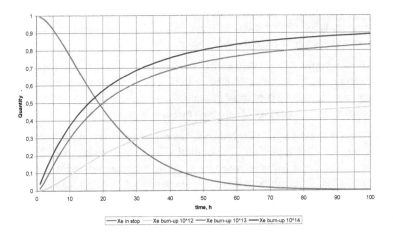

Figure 5. Dependence of ratio neutron absorption in ^{135}Xe to ^{135}I formation on fuel work time and neutron flux.

Data from charts on figure 5 can be used for estimation of ^{135}Xe neutron absorption at dynamic loading regime. For example, in neutron flux $5*10^{13}$sm^{-2}s^{-1} with 5 work hours and exposition 45 hours neutron absorption in ^{135}Xe is less than 30% of fission product ^{135}I formation and is about ~1.9 %. Saved 4 % of neutrons can be used for construction material absorption and leakage.

For effective usage of dynamic loading duration of working regime can be in the region of 5 to 10 hours and exposition time – 35-50 hours. Increasing exposition time more than 60 hours is unreasonable. So fuel mass can be 3.5 – 5.0 times larger.

Dynamic loading regime for traditional fuel types with rigid placed fuel assemblies in a core is sufficiently complicated. In general it can be made in reactors, which have fuel assemblies' replacements without reactor shut down, such as CANDU or RBMK-1000. But large amount of replacements, needed for this regime realization, decreases durability of these fuel assemblies and replacement mechanism.

Dynamic loading regime is possible in molted salt reactors [8] and in reactors with spherical fuel circulation in heat-exchange loop [9]. Work [10] shows that this regime can significantly simplify a molted salt reactor technology of fuel purification from fission products and actinides.

6. Joint work of zones with different campaign moments in the same core

6.1. Ideal reactor with zero loss of neutrons

Modern power thermal neutron reactors widely use campaigns with multiple fuel reloading [11, 12]. During each reloading fuel with maximal burn-up is moved out core, fuel with different burn-up is rearranged and fresh fuel is loaded. Necessity of reloading is caused by fission materials concentration decrease during irradiation.

It is shown in presented above data that high concentration of fission materials is stabilized in high fission materials reproduction reactors during campaign burn-up. Burn-up increase in campaign of such reactors is purpose of multiple fuel replacements with entering regime of negative reactivity in some part of core with maximal burn-up.

Terms of detailed campaign and compact campaign are introduced here for better understanding. Detailed campaign is theoretically calculated campaign with continuous reactor work with use of negative reactivity. Compact campaign contains the portions of various fuels with different burn-up level. Durability of compact campaign is equal durability of detailed campaign divided on number of fuel zones with different campaign burn-up. For work regime with fuel replacements is used term "zone superposition".

It can be noted that work in compact regime is possible only if neutron absorption in control elements in detailed campaign is more than zero.

Fuel with nuclides ^{235}U and ^{238}U in zone superposition regime is considered. Figure 6 shows detailed campaign characteristics of initial fuel ^{235}U (0.47 %) + ^{238}U in neutron flux $9*10^{13}$ sm^{-2}s^{-1} and multiplication factor of compact campaign with four fuel replacements. In superposition regime fuel works in reactor for 25000 hours. Raw uranium usage in campaign increases up to 3.81%. Multiplication coefficient at the end of detailed campaign becomes lower than unity by 0.04.

Characteristics of campaign with the same fuel type and different initial contents of fission materials in the beginning of campaign are presented at the string 10 of Attachment Table 1. Here it is corresponds to natural uranium fission materials contents. Larger burn-up and raw uranium usage in open fuel cycle is obtained in this variant.

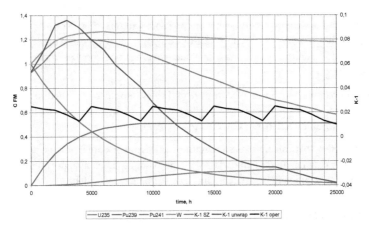

Figure 6. Detailed campaign characteristics with initial fuel ^{235}U + ^{238}U and multiplication factor of compact campaign with four fuel replacements (string 9 of Attachment Table 1).

6.2. Reactors with non-zero loss of neutrons

The string 11 of Attachment Table 1 shows characteristics of detailed campaign with fuel ^{239}Pu + ^{239}Pu + ^{238}U and multiplication factor of compact campaign with eight fuel replacements. Neutron loss in construction materials and leakage is 1.7% in this campaign.

This campaign is identical to the campaign, which is presented at the string 2 of Attachment Table 1 and figure 3, by the fuel type and composition. In spite of neutron loss presence larger campaign duration and burn-up is obtained in the campaign.

Comparison of campaign characteristics with natural uranium and different neutrons loss in the construction materials and leakage is presented at the strings 10, 12, 13, and 14 of Attachment Table 1. Selected data from this table is presented in Table 2.

1	Neutron loss, %	0	1.7	2.8	5.2
2	The duration of the campaign, hour	34 000	29 000	24 000	19 000
3	Burn-up, %	3,92	3,38	2,84	2,41

Table 2. Comparison of campaign characteristics with natural uranium and different neutron loss in zone superposition regime.

Characteristics of reactor campaign fueled with natural uranium and neutron loss 5.2% is presented at the string 14 of Attachment Table 1, and at the string 12 of Attachment Table 1– campaign with the same fuel and neutron loss 1.7%.

^{241}Pu production in reactor campaign with 5.2 % neutron loss is only reaching stationary level. Stationary level of ^{241}Pu production in the campaign with 1.7 % neutron loss is reached.

For comparison, characteristics of reactor campaign with the same fuel and 5.2 % neutron loss but without zone superposition using are presented at the string 23 of Attachment Table 1. Such campaign has the worst characteristics among others by campaign duration, fuel burn-up and raw uranium usage, and Rsz value – average neutron loss in control elements during campaign.

In reality campaign at string 23 of Attachment Table 1should be finished at least at 1000 hours earlier on condition of sufficient positive reactivity. Characteristics of this campaign are quite close to characteristics of CANDU reactor campaign.

It should be noted, that calculations shown in the present work are made for point model of reactor. In cases, when fission materials contents change at campaigns is minimal, real reactor campaign characteristics matches this calculation in great extent.

7. Fission materials production in control elements

7.1. Task formulation

Previous chapter material analysis shows that campaign characteristics improving, including the reproduction coefficient, with the same fuel type is obtained by decrease of neutron absorption in control elements. Improvement not always can be made by neutron flux change or superposition regime implementation. It is interesting to replace useless neutron absorption in control elements (cadmium, boron, gadolinium etc.) by absorption in raw fuel materials (^{238}U and ^{232}Th). First activation products of raw nuclides must be deleted from the core. This technology should not permit fission materials formation in chain of product transformation in the core. Activation chains of ^{238}U and ^{232}Th are:

$$^{238}\text{U} + \text{n} \rightarrow ^{239}\text{U} \left(1412 \text{ s}\right) \rightarrow ^{239}\text{Np} \left(2.33 \text{ d}\right) \rightarrow ^{239}\text{Pu}; \qquad (8)$$

$$^{232}\text{Th} + \text{n} \rightarrow ^{233}\text{Th} \left(1325 \text{ s}\right) \rightarrow ^{233}\text{Pa} \left(27.4 \text{ d}\right) \rightarrow ^{233}\text{U}; \qquad (9)$$

Uranium chain has the minor time reserve, but it is big enough.

Any technology always supplies some part of final fission product into the absorber placing area. It is simple to estimate the negative effect of such penetration. It is defined by purification time outside core from primary activation products (^{239}Np and ^{233}Pa) and degree of fission materials penetration into core loop.

It can be noticed that equal effect from absorption in raw and fission nuclides of uranium chain is obtained at raw to fission nuclides contents ratio 530, and for thorium chain it is equal to 182. Comparison of fission nuclide half-life periods, ratios of raw and fission components with same influence to reactivity, and fission characteristics (^{233}U has larger part of fissions from total neutron absorptions) shows, that thorium is preferable primary candidate for control element development with fission material production. Estimation effectiveness can be made by division of precursor half-time to ratio of raw and fission components with same influence to reactivity. For thorium this parameter is 3.6 hours, and for uranium – 6.33 minutes. Difference is 36 times.

This suggestion realization does not mean that all control elements must produce fission components. There is no need to have it in safety system with total neutron absorption at zero level in any case of campaign.

At initial development stage it is possible to divide two functions of control system: traditional, for fast regulation of reactor power with efficiency of 1 β and slow regulation system with fission materials production and efficiency of maximal reactivity in campaign.

Part of neutrons involved in fission materials production is linked with produced fission material amount and energy emitted in core by formula:

$$n = Q * g * \xi; \tag{10}$$

Where:

n – amount of produced fission materials, nuclei;
Q – fission energy emitted in core, J;
g – link coefficient between emitted power and fissions number ($3.1*10^{10}$), J^{-1};
ξ – part of neutrons, which are involved into fission materials production.

With core power 1000 MW and 1 % of involved in fission materials production neutrons amount of produced fission material is equal to $1.2*10^{-4}$ g per second. For year mass is ~3800 g.

It is possible to use value of fuel burn-up and number of fuel reloading during the campaign. Insertion of produced fission materials is made with each new fuel portion. In this case the number of inserted nuclei of produced fission materials is described by formula:

$$n1 = m * \chi / j; \tag{11}$$

Where:

n1 – amount of nuclei of produced fission materials, which are inserted with each fuel reload;
m – amount of fuel nuclei, including raw and fission components;
χ – fuel burn-up during campaign;
j – number of fuel reloads during campaign.

Calculation of campaign characteristics with fission materials production by use of redundant reactivity compensation is made by iterative method. Adsorption of redundant neutrons at this process leads to additional reactivity insertion. At the good realization of campaign this additional reactivity is added to areas of detailed campaign with negative reactivity, which supplies prolongation of initial fuel work duration. Iterations in carried out calculations are not always optimal. Measures of optimal iteration are values of R_{min} and R SZ, which are shown in the Attachment Table 1.

7.2. Fuel is natural uranium

Let's examine how neutron loss in control elements used for generation of ^{235}U in reactor with initial fuel ^{235}U + ^{238}U influence on campaign characteristics (figures 10, 12 - 14). Characteristics of detailed campaign with 10^{14} sm^{-2}sec^{-1} neutron flux, ^{233}U production and 1,7 % neutron loss are shown at the string 15 of Attachment Table 1.

Maximum fuel burn-up (5.27%) is reached in this campaign in comparison with previous campaigns; duration of detailed campaign is 34000 hours. Reactivity change during work is close to optimum results (K-1 from 1 % up to 3.2 %). Reactor power is stable. Reactor power change in constant neutron flux is not exceeding 2.5% from the average. ^{233}U generation increase in the campaign with this fuel type is possible at the expense of operating reactivity pike lowering.

Enough good indexes by this technology can be received into the reactor with 5.2% neutron loss. Description of such campaign is presented in string 24 of Attachment Table 1. Indexes of such campaign are high enough.

8. Characteristics of reactors with high fission materials reproduction

8.1. Possibility of neutron loss decrease

Obtaining of high fission materials reproduction is possible only in reactors, which have the minimal neutron loss in construction materials and leakage. This loss must be about 1-2% by the preliminary analysis of previous materials. It is not possible to realize on practice all of the discussed fuel cycles.

Let us consider in this chapter could be achieved good results with bigger neutron loss level and what must be considered as good results.

CANDU (and its many versions) with heavy water as coolant and moderator, with zirconium shells and channel walls is the nearest reactor to high fission materials reproduction reactors. Let's look at the characteristics of this reactor and its possible modifications, which are directed to neutron loss decrease. Fuel assembly of CANDU reactor with 37 fuel element (on left) and modification of this assembly with replacement of 7 central fuel element by beryllium block are presented in the figure 7.

Characteristics of initial reactor and its 6 modifications are presented at the table 3:

1. Reactor with fuel assembly at the figure 19 (on the left) with standard reflector, standard fuel on the basis of natural uranium dioxide;
2. Reactor by the i.1, with enriched by isotopes ^{90}Zr and ^{120}Sn in fuel rod shells and fuel assembly casing zirconium and tin.
3. Reactor by the i.2, with bigger thickness of heavy-water reflector.
4. Reactor by the i.2, with addition of graphite reflector to heavy-water reflector.
5. Reactor by the i.4, with fuel rods made of metallic uranium (80% of volume) and bismuth (20% of volume).

6. Reactor by the i.5, with fuel assembly shown at the figure 19 (on the right) with beryllium insertion.
7. Reactor by the i.6, which differs by using of heavy water instead of graphite layer in reflector.

№	Parameter\variant	1	2	3	4	5	6	7
1	Leakage	1.7904	1.7789	1.1121	1.1354	0.915	1.041	1.182
2	D	0,163	0.172	0.391	0.221	0.147	0.213	0.254
3	Be						- 0,513	- 0,531
4	C				0.356	0,285	0,327	
5	O	0,443	0,441	0,489	0,456	0,396	0,399	0,410
6	Zr	2,54						
7	90Zr		0.281	0.283	0.281	0.236	0.240	0.239
8	Sn	0,125						
9	120Sn		0.0298	0.0287	0.0319	0.023	0.024	0.0254
10	Bi					0.079	0.078	0.0772
11	Σ	5,061	2,70	2,30	2,481	2,081	1,809	1,660
12	K eff	1,1171	1.1473	1.15489	1.15093	1.1265	1.13233	1.13428
13	RC	0,808	0,804	0,80	0.802	0,871	0,855	0,853
	Reflector	D2O 21		D2O 62		D2O+C		D2O 62
	Cell	37 fuel rods					30 fuel rods + Be	
	Fuel	UO2				0.8*U + 0.2*Bi		

Table 3. Neutron loss (string 1-11, %), multiplication factor and reproduction coefficient (string 12, 13, relative units) and fuel types and reflector types in variants of CANDU reactor models.

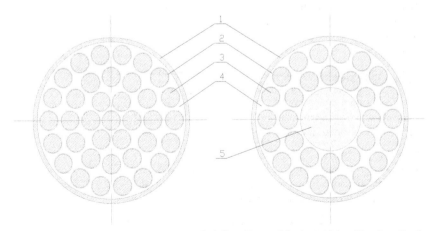

Figure 7. Fuel assembly of CANDU reactor (on the left) and its modification with beryllium insertion (on the right). 1 – fuel assembly casing; 2 – fuel element shell; 3 – fuel rods; 4 – coolant; 5 – beryllium insertion.

Analysis of calculation results allows saying:

1. Initial variant of reactor has the worst characteristics;
2. Major change of properties is reached by the change of natural zirconium and tin to its isotopes ^{90}Zr и ^{120}Sn. These isotopes have the best properties among the rest of isotopes of these elements and have the considerable contents in natural elements. Neutrons loss of this changing has decrease till 5.06% to 2.70%.
3. Decreasing of leakage at the expense of reflector thickness increasing at the i.3 and 4 also leads to appreciable lowering of neutrons loss.
4. Important improvement of characteristics is obtained by replacement of oxide fuel by metallic, including the reproduction of fission materials.
5. The best parameters among the presented variants have the reactor, with fuel assembly made from 30 fuel rods and beryllium insertion. Differences between 6 and 7 variants, which is differ by reflector construction, are minor.

1.7 %, 2.8 % and 5.2 % neutron loss are used in Table 1 as results of CANDU reactor calculation.

8.2. Construction of fuel assembly with composite core

Good neutron-physical characteristics of metallic fuel are well-known [13, 14]. Such fuel is used on reactors by the first stages of atomic energetic progress. Significant swelling under the enough small burn-up is the lack of it. Maximum allowed burn-up of "KC-150" [15] reactor is settled by the level in 15 MW*day/kg. It is indicated that this burn-up equal to 4 MW*day/kg the maximal extension of fuel element is laying down for 5-7% under the temperature to 350⁰C. Rod-shaped fuel element of traditional construction with such characteristics of swelling cannot be perspective for reactors with burn-up of 40 MW*day/kg and higher.

Situation can be changed, if we will use the fuel element with composite metal core. 2 variants of such fuel element construction are presented at the figure 8. Core of such fuel element is containing fuel elements 2 and 3, liquid-metal filler 4, which are placed under the protection cover.

In variant, which is placed on the left, the core contains 8 leaf fuel cells and one cylindrical fuel element. These elements at swelling occupy the space, which was filled by liquid metal (with fuel working temperature). At the second variant only 2 fuel elements are used. These elements are identical and inserted into each other. Cuttings, which bring down the pressure during swelling in radial direction, are made on elements surfaces.

Bismuth, lead [16] and tin can be filler material. If cover is made from zirconium, then lead, by preliminary estimations, leaves the list of candidates because it actively interacts with zirconium and breaks its initial structure. It is possible, that bismuth in clean type will actively interact with zirconium with such result.

Tin at the time of interaction with zirconium forms the solid compound – zirconium stannide Zr_3Sn_2 on its surface, which is melted at temperature 1985 ⁰C [17]. Tin is included into the composition of zircaloy, which is used in fuel element covers. Shortage of tin (its enough high absorption cross-section) is possible to delete by using tin enriched by ^{120}Sn isotope. ^{120}Sn has the best properties among tin isotopes with even atomic weight.

It is possible, that alloy of tin with bismuth and lead also will form the tin stannide on the surface of zirconium, which prevents the interaction of zirconium with bismuth and lead. Rigorous research in this direction is necessary.

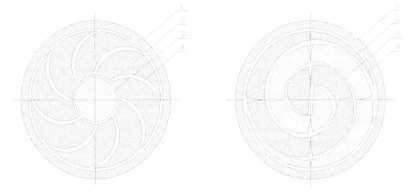

1 – fuel element cover; 2 and 3 – fuel elements; 4 – filler.

Figure 8. Variants of fuel element construction with composite metallic core.

We should specify requests to whole of possible varieties of core composite elements, which must have high workability of fuel element with maximal level of burn-up.

In this case the statement is follows: space between particles must create stable conditions for separate particles location under fuel element cover when increasing volume in specified limits without appearing of additional effort between particles;

This claim dissects for some independent claims:

- Allowed reduction of areas gage, which is occupied by composite core, is settled;
- In the initial condition surface of particles is must be distant from each other by the all surface of particles;
- Crush elements might be placed between particles without changing the distance between them for realization of increasing of particle sizes;
- In the working progress, under the decreasing of crushed elements dimensions it is desirable, that the distance between particles centers is staying invariable or changing negligible.

It should be marked that important factor in such construction of fuel element, which decrease the metallic fuel elements swelling, is their small thickness. The small thickness increases the migration of fission gaseous splinters over the external surface with decreasing of inner mechanical stresses.

By the estimations, forms changing of metallic fuel rods under the execution of given claims and initial content of liquid-metal filler in 20% by the volume is not leading to trespassing of fuel element cover under the burn-up till 40-59 MW*day/kg. Such fuel has being used in calculations of reactor models, which was presented in the table 1.

8.3. Open fuel cycle

Open fuel cycle can be realized by the initial fuel, which consists of ^{235}U and ^{238}U. This fuel can contain natural uranium, as in CANDU reactors, enriched uranium, as in the majority of modern reactors, and, as it presented on the calculation above, from the ^{235}U impoverished. Fission materials in all of these cases are made from natural uranium passing the chains, which is bounded with recycling of spent fuel. Let's assume that technology of ^{233}U generation is not bounded with recycling of spent fuel and might take the part in opened fuel cycle.

The final product of open fuel cycle can be the perfect base or composite part of closed fuel cycle. Especially, if fission isotopes of plutonium come to equilibrium.

Open cycles, which are presented in strings 1, 9, 10, 12 -15, 17, 23 and 24 of Attachment Table 1, are described above. Maximum portion of natural uranium usage on it is reached under the conduction of superposition mode areas with ^{233}U generation. Quite high part of using is reached in the superposition mode under the decreased neutrons losses in construction materials and for leakage. In these cases fission isotopes of plutonium are in equilibrium at campaign end. Maximal reached part of natural uranium using in presented cycles forms 5.27%. This amount in cycles with generation of ^{233}U can be increased at the expense of generation mode optimization. Optimization is not appeared by paramount task at this work.

Thereby, there are different ways for reaching of high natural uranium usage in open fuel cycles and reception of closed fuel cycles fission components on its.

8.4. Closed fuel cycles.

In this work it is considered that all initial fuel products use technology of spent fuel recycling. In closed cycles it is possible to use raw uranium based fuel or raw thorium based fuel.

8.4.1. Uranium-plutonium fuel

The best closed fuel cycle characteristics are achieved with use of equilibrium contents of fission materials. But the ideal equilibrium uranium-plutonium cycle has very small contents of fission materials so the neutron leakage plays more important role than in CANDU with raw uranium.

Fission materials contents in equilibrium uranium-plutonium fuel equals 0.34 % in condition of resonant neutrons absorption absence in ^{238}U (in CANDU its contents is more than twice larger – 0.712 %).

Solution of the task is possible in two ways – 1) by increasing of initial contents of ^{235}U by adding it, 2) increasing of fission materials contents by increasing resonant absorption in ^{238}U.

The string 19 of Attachment Table 1shows variant with addition of 0.35 % [235]U from mass of [238]U, that equals equilibrium contents of [239]Pu and [241]Pu. This fuel can be made by plutonium isotopes, which are extracted from spent fuel, addition to a mix of raw uranium (half of [238]U mass) and [238]U from spent fuel.

The string 20 of Attachment Table 1 shows campaign characteristics with initial fuel containing equilibrium contents of [239]Pu and [241]Pu and addition of [235]U, which contents is much less than in previous fuel. Resonant neutron absorption in [238]U is 1.2 times higher. Production of [233]U is lead. Equilibrium contents of [239]Pu and [241]Pu is also 1.2 times greater.

Burn-up of 3.49 % and raw uranium usage 12.12 % is achieved.

String 18 of Table 1 of Attachment shows campaign with this technology. Strings 21 and 22 show campaign with no [233]U production. Maximal burn-up 3.51 % and maximal raw uranium usage 14.28 % is achieved.

8.4.2. Uranium-thorium fuel

Equilibrium cycle with uranium-thorium fuel even without resonant neutron absorption in thorium has high fission materials contents, which mean possibility of high burn-up achieving in detailed campaign. Features of uranium-thorium campaign are high neutron absorption in [233]Pa and its long half-life, which are used in technology of [233]U production by reactivity margin decreasing.

Detailed campaign with neutron flux about 10^{14} sm^{-2}s^{-1} becomes very short even in case of low neutron losses which equals 1.7 %. The campaign with twice lower neutron flux is possible at neutron losses of 5.0 %.

Figure 9 shows detailed campaign characteristics with equilibrium uranium-thorium fuel with [233]U production at neutron flux $6*10^{14}$ sm^{-2}s^{-1} and neutron losses 5.0 %.

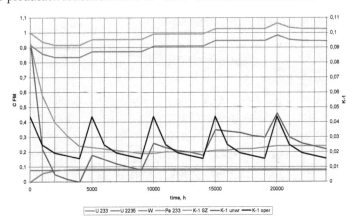

Figure 9. Detailed campaign characteristics with equilibrium uranium-thorium fuel and [233]U production. Neutron flux is $6*10^{14}$ sm^{-2}s^{-1}, neutron loss – 5.0 % (string 25 of Table 1 of Attachment).

High stability of reactor power is obtained in compact campaign. Reactivity margin decrease in detailed campaign is observed in first 5000 hours. During all time of detailed campaign reactivity stays positive. Reactor work prolongation over 24000 hours is possible. At the shown campaign burn-up of 4.35% is reached.

The difference of the campaign from uranium-plutonium campaign is possibility of significant amount increase of fission materials to the end of the campaign. Reproduction coefficient of the campaign, which is shown at figure 20, is more than unity by 11 %.

8.4.3. Mixed fuel

Mixed fuel (with uranium-thorium raw materials) is rational to use in closed cycles with equilibrium between fission and raw materials.

This fuel is less dependent from neutron flux than pure thorium fuel. Together with it the fuel is more sensitive to the neutron losses. At ^{238}U contents 75 % in campaign beginning, fission materials contents 1.16 % neutron losses 2.8% neutron flux less than $8*10^{13}$ $sm^{-2}s^{-1}$ is possible effective work. The duration of the campaign can be 34000 hours.

Campaign with mixed fuel can supply simultaneous use of thorium and uranium fuel with burn-up corresponding to its natural contents with high and even full use of raw uranium [18].

Closed fuel cycles with uranium fuel after several cycles can have increased contents of ^{242}Pu, which is feeble burning out and being absorber and filler decreasing of campaign effectiveness. The possible way of solving is plutonium isotope separation. In mixed fuel ^{242}Pu accumulation will be less and its critical contents will be reached much later. The losses for it are also less.

8.5. About neutron balance in the campaign

In previous materials model of abstract reactor with parameter "neutron losses in constructive materials and leakage" is used. Dependence of reactor characteristics and its campaign is more complex in real life.

More detailed presentation of these relations can be obtained from neutron balance during reactor work. Neutron creation is present not only in fission nuclides but in raw fuel components and such constructive materials as heavy water and beryllium.

Figure 11 show neutron balance in reactors with two types of fuel assemblies with different coolants. Fuel assembly for heavy water coolant is shown at figure 10-a, and fuel assembly for liquid metal – at figure 10- b. Measures for neutron energy loss prevention at fuel location are used in fuel assembly for liquid metal coolant.

Reactor calculations with use of program [19] are conducted, which are the base for reactor campaign calculations with use of program [20].

1 – case of fuel assembly, 2 – gaseous gap, 3 – screen, 4 – coolant, 5 – fuel rod, 6 – beryllium insert for a) and gaseous cavity for b).

Figure 10. Fuel assemblies with water (a) and liquid metal (b) coolant.

Neutron absorptions (black columns) neutron fissions (blue columns) in different fuel nuclides are shown on diagrams. Two types are used ^{235}U – natural raw uranium signed as U235N, and formed during transformations of thorium chain (^{232}Th-^{233}Pa-^{233}U-^{234}U-^{235}U), signed as U235S.

Fission nuclides have columns for difference between secondary neutrons and total neutron absorptions (red columns). Difference for raw ^{232}Th and ^{238}U between number of secondary neutrons and number of total absorptions with fission are indicated in yellow columns. Columns 1 and 2 indicate neutron absorptions in construction materials and fission products correspondingly.

In left part of each column lay data for reactor with liquid metal coolant, in right part – data for reactor with water coolant/

Total height of red and yellow columns must be equal to total height of black columns for nuclides ^{232}Th, ^{238}U, ^{234}U, ^{240}Pu and columns for neutron absorptions in construction materials and fission products.

If neutron loss in construction materials and leakage are less, than more neutrons can be absorbed in fission products, which number increases during campaign.

For the compared variants fission activity in raw nuclides and ratio of absorptions and fissions in ^{241}Pu are better. In result total neutron absorption in fission products (8 %) for heavy water coolant reactor is less than total neutron absorption in fission products (13 %) for liquid metal coolant reactor at the same neutron losses for construction materials and leakage. So reactor with liquid metal coolant has campaign with burn-up 25.5 MW*day/kg when heavy water coolant reactor has campaign with burn-up 11,3 MW*day/kg. Liquid metal cooled reactor has higher raw uranium usage in open cycle campaign.

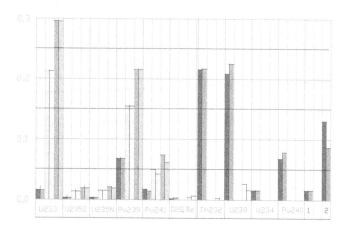

Figure 11. Liquid metal and heavy water cooled reactors neutron balance in campaign with equilibrium by fuel nuclides regime.

8.6. External neutron sources for reproduction coefficient increase in thermal reactor

External neutron sources can be included in neutron balance if it is linked somehow with reactor work. There are schemes, where neutrons, which are formed in reactions between accelerated protons and nuclei of heavy metals, are emitted to the reactor's core. This scheme has significant lack of reactor work reliability because of use less reliable source – electric nuclear installation. When autonomic electric nuclear installation is used only for [233]U formation by neutron absorptions in thorium, reactor work almost is not depending on work reliability of electric nuclear installation. Work [21] demonstrates that raw uranium usage can be increased up to 100 % with use of comparatively small energy expenses (less than 2,5 % of produced energy).

Note, that modern thermal reactors, which use less than 1% or raw uranium, about 1% of produced energy spent for fuel enrichment.

Possible characteristics of joint work of fission reactor and neutron source based on fusion reactor are shown [22].

9. About development directions of nuclear reactors and devices for energy transformation

Achieving of good results of raw uranium usage with small uranium mass used cannot be the sole claim for highly effective reactor. Claim for effective energy transformation is also significant. High value of efficiency allows not only decrease uranium usage for unit of power production but decrease of negative effect to the nature, because small efficiency means more thermal energy release.

Complexity of nuclear reactor technology, effective work with energy transformation installations requests contradictions make us to look for different technical solutions, which can provide desired result. During development of atomic power plants many solutions, different coolant types, construction materials, schemes of thermal to electrical power transformation were tested.

In field of thermal reactors the base niche is occupied by water-water reactors. Between two types of these reactors (pressurized and boiling) there is a competitive contest. Now in heavy water moderated reactors there are both transfer directions of water coolant thermal energy to turbine. Maximal efficiency of this scheme is limited by ~34 %, which is depending on acceptable water pressure in a core and its temperature correspondingly.

High efficiency can be obtained in fast liquid metal cooled reactors with coolant temperature about 500 °C, which is typical for combustible fuel powered electro stations. Feature of this scheme is decoupling of coolant load conditions and Rankin cycle actuating medium. Pressure of coolant in reactor vessel can be close to atmospheric.

Decoupling of coolant load conditions and actuating medium also exists in schemes with gaseous coolant [23]. Energy transfer direction with use of gaseous coolant in 50-60-ties of XX century has transformed by type of energy receiver from Rankin cycle to Briton cycle [24].

Potential some efficiency increase (up to ~48 %) in this scheme is possible, but has some negative effects. Main effects are listed below.

The first. There is a need to steep increase of coolant temperature, up to 1000 °C. It causes increase of expenses for reactor and energy transformation installation.

The second is fuel rod design complication caused by higher coolant temperature. Besides expenses increase it causes tendency of fission products release increase.

The third is power increase on the Briton shafting four times higher than Rankin steam turbine with the same power output and cost rise, correspondingly.

All these factors are negative for NPP economy, time for solution preparation to practical realization.

The paper [25] shows usage possibility in Briton cycle instead of turbines piston machines, which supply high efficiency obtaining possibility at lower coolant temperature. It is obtained because of absence energy transfer chain from high speed gas to blades of turbine (and vice versa in compressor), which makes basic energy loss in Briton cycle. Shortages of this solution are absence of practical schemes of needed piston machines and less power of a unit.

9.1. Heavy water reactor with gaseous coolant in Rankin cycle

9.1.1. Technology problems

In channel heavy water reactors portion of energy, which is released in moderator, is lost for energy production because the moderator has low temperature. In channel released energy is used for energy transformation. Together with heat flow to moderator from channel thermal energy loss is about 10 % from total energy released in the core. Maybe it is a reason of HTGR design with high temperature graphite, which transfers energy of neutron moderation to fuel assemblies.

Rankin cycle with two steam overheating, one at maximal pressure in cycle (at the entrance to turbine) and additional – after steam expansion to specified pressure, is used in modern reactors. At small maximal pressure levels steam at last stage and turbine exit is wet with high specific humidity. It leads to exit blades corrosion, necessity of valuable alloys usage. Blades of last stage have maximal size and determine total cost of turbine.

9.1.2. Solution description

Scheme of joint work of heavy water reactor with gaseous coolant and Rankin cycle steam turbine, which is based on full use of fission energy (including neutron moderation energy) and three stepped steam entering to turbine, is suggested in the paper [26].

First feature of the solution is based on heat emission to the steam in Rankin cycle, which is made not in a single process, but several with different temperature intervals.

The second feature allows avoiding presence of steam with high specific humidity on exit stage blades of turbine. Good parameters are achieved at maximal steam temperature 500 °C and maximal pressure 20.0 MPa.

Scheme of coolant ducts and steam loop of heavy water channel reactor with gaseous coolant is shown at figure 12. Differences of this scheme from other solutions are usage of neutron moderation energy in Rankin cycle for water of steam loop heating, separation of coolant duct on four ducts, which supply with energy re-heater and water evaporator, separate steam overheaters. Coolant ducts for heating and water evaporation can be water ducts.

Neutron moderation energy transfer to water in Rankin cycle is realized with use of vessel of reactor design with greater than atmospheric pressure, but less than maximal pressure in coolant duct. It allows having moderator with temperature greater than water heated in Rankin cycle and decrease thickness of duct walls.

Separation of overheaters ducts allows decreasing of danger of hermetization loss of steam loop with high pressure. Actions for sequence avoiding of hermetization loss is made to duct with small portion of reactor power (~20 %). Reactor scheme at figure 12 gaseous coolant duct, which is linked to high pressure steam duct, has vessel, which supplies decrease of maximal pressure at accidental steam leakage to coolant (volume of steam is limited) and increase time of pressure growth in coolant duct at such leakage (work condition of automatics becomes better).

Besides this separation optimizes energy expenses for coolant pumping by use of temperature differential in corresponding external heat exchangers.

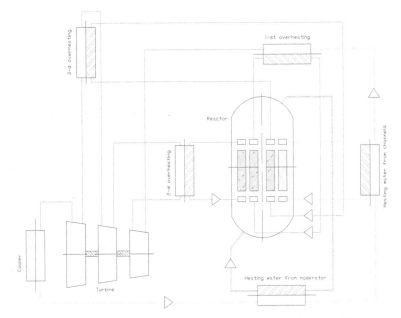

Figure 12. Scheme of coolant ducts and steam loop of heavy water channel reactor with gaseous coolant.

9.2. Description of reactor with thermal power 80 MW

Calculation of neutron and thermal characteristics of the reactor are conducted with use of programs [16, 27], campaign characteristics with use of program [19], Rankin cycle and turbine characteristics with use of programs [28, 29].

Sketches of reactor design with thermal power 80 MW with gaseous cooled fuel assembly and with water cooled fuel assembly are worked out. Water cooled fuel assemblies are located in central part of the core.

The reactor has 85 fuel assemblies, which are located in nodes of triangular grid with step 18 sm. Core is surrounded by two-layer reflector. Inner layer is heavy water, outer layer is graphite.

9.2.1. Description of fuel assembly

Each fuel assembly has 59 fuel rods with outer diameter 6 mm.

Coolant cross-section is limited by a screen made of zirconium alloy thin shell (thickness 1 mm), and a casing made of the same alloy shell (thickness 3 mm) at distance of 2 mm from the screen. The gap between the screen and the casing is gas filled. The gas pressure equals to average by channel height pressure of actuating medium.

Gaseous coolant in fuel assembly can be hydrogen or helium. Fuel on base of metallic uranium and thorium is chosen for the work variant (figure 13). Each fuel rod has uranium fuel elements alternating by height with thorium fuel elements [30]. Initial contents of ^{235}U in uranium elements is 0.5 %. Initial contents of ^{239}Pu and ^{241}Pu in uranium elements, and initial contents of ^{233}U in thorium elements equal to equilibrium contents of these nuclides during durable campaign.

This design supplies constant energy release distribution by height of fuel rod independent from fuel type under the shell – uranium or thorium.

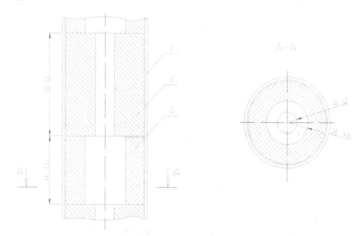

1 – rod shell, 2 – uranium fuel element, 3 – thorium fuel element.

Figure 13. Disposition of uranium and thorium fuel elements by fuel rod height.

Quality of fuel rod height energy release distribution becomes better with increase of portion of equilibrium nuclides and, correspondingly, raw uranium usage portion increase.

Difference of fuel elements form with uranium and thorium has significant role at reprocessing of spent fuel, when there is need to separate uranium and thorium fuel.

With use of equivalent fuel elements with mix of uranium and thorium, where also can be obtained uniform energy release, we will have problem of separation fission ^{233}U and ^{235}U from raw ^{238}U. Without this separation raw uranium usage portion has steep decrease.

9.2.2. Campaign characteristics

Change of reactivity margin during two variants of the campaign without fuel replacement and with one fuel replacement.

At constant neutron flux in core ratio of reactor power at campaign end to power in campaign beginning in variant without fuel replacement equals 37,6 %, and with one replacement – 17,6%. Fuel burn-up is 5,1 % or 48,7 MW*day/kg.

Uranium tablets of first campaign are made from raw uranium with adding of 0,357 % ^{239}Pu and 0,12 % ^{241}Pu. It is possible to have the same mass of nuclides with different mixture contents. In thorium tablets of the first campaign is added 2.3 % ^{235}U of high enrichment.

Uranium tablets are divided on two parts at spent fuel reprocessing. Fission products and plutonium is extracted from 70% of spent fuel uranium mass of the first campaign. This fuel part is replaced with raw uranium, in which extracted plutonium is added.

Fission products are separated from the second fuel part. Add mass of raw uranium which equals mass of burned ^{238}U in uranium fuel.

Fission products are separated from the thorium fuel part. Add mass of thorium which equals mass of extracted fission products.

Usage of raw uranium in closed cycle is 5%.

In reactor core there is 2312 kg of fuel. Spent fuel reprocessing requirement is 680 kg per annum. Raw uranium requirement is 510 kg per annum. Required mass of thorium is 14 kg per annum.

For comparison, reactor WWER-1000 uranium requirement is 10 times larger per unit power and relative amount of dissipated power is more 1.4 times.

9.2.3. Thermal characteristics of fuel assemblies

Thermal characteristics of fuel assemblies with hydrogen coolant are presented in table 4. Characteristics of fuel assemblies with helium coolant are slightly different. Hydrogen is cheaper than helium so it is preferable.

Coolant	H_2			
Gas in the gap of fuel assembly	CO_2			
Function of fuel assembly	3 overheating	2 overheating	1 overheating	Water heating
Temperature at fuel assembly entrance, ^0C	285.9	285.3	376	141
Temperature at fuel assembly exit, ^0C	510	510	510	375
Fuel assembly number in reactor	10	12	17	46
Coolant flow in fuel assembly, kg/s	0.286	0.285	0.478	0.274
Coolant velocity at fuel assembly entrance, m/s	28.8	28.6	55.36	20
Coolant velocity at fuel assembly exit, m/s	40.1	40	67.35	31.5
Pressure difference in fuel assembly, Pa	3892	3882	9617	2923
Coolant pump power in fuel assembly, kW	5.24	6.25	36.8	17.34
Total coolant pump power, kW / % from total	65.63/0,2			

Tmax at fuel rod shell, °C	805		758	813
Tmax fuel rod, °C	1125		979	1035
Tmax of gas gap of fuel assembly, °C	337	336	388	203
Tmax of fuel assembly screen, °C	364		420	223
Tmax of fuel assembly casing, °C	182	181	202	168

Table 4. Thermal characteristics of fuel assemblies with gaseous coolant.

Basic features of these assemblies are listed below.

- Work of all fuel assembly elements is conducted in acceptable temperature region.
- Hydraulic loss for pumping is small.

All this features obtained with simple fuel assembly design and its base elements – fuel rods. Used fuel rods are well fine-tuned at practice of large number of reactors. Small hydraulic losses lead to small required power of coolant pumps. It characterizes costs level for reactor creation and losses level during exploitation.

9.2.4. Rankin cycle characteristics

Calculations of cycle variants, which are different by maximal steam pressure, number of interim stages of re-heating and steam expansion characteristics on it, are conducted. The most economic variant of possible by complexity manufacturing is shown at table 5 and figure 14.

This cycle use two interim overheating and minimal total steam expansion. Zero water content in steam at turbine exit is obtained. This means small turbine HWD, usage possibility in turbine design cheaper materials for the blades and correspondingly smaller cost.

Cooler of this variant also has small HWD and comparatively high temperature of fluidized steam supplies possibility of its use as heat source for home and business needs.

The second variant has steam quality on the exit of turbine 0.93, which differs much from steam quality in VVER – 0.75. Total steam expansion is 3.7 times bigger than in VVER, and 42.4% efficiency is obtained.

Problem of material for turbine last stage blades here is not so strong, as in nuclear power plants with VVER or ABWR reactors [31].

For electro stations, which work in autonomic regime, for example in distant areas from common electrical supply network, the first variant is preferable by two reasons:

1. In this case obvious, that there is need in heat of low potential;
2. Electro station cost is significantly cheaper with less HWD of turbine and less cost of blades. Turbine cost is significant part of NPP cost [31].

№ re-heater	T, °C	P, gauge atmospheres	Enthalpy	Portion of Q	Entropy	Sp. volume	Vapor quality
1 Cooler	69,6	0,304	2625,4	-0,583	7,764	5,17	1
	69,3		290,1		0,946		
2 Water from moderator		10		0,08	1,645	~0,001	0
	131		610,5		1,626		
3 Water from fuel assembly	365		1811,5	0,2998	3,99		
4 Water-steam		200		0,1527			
	366		2423,1		4,95	0,0074	
5 The 1-st overheating	500		3241,2	0,2042	6,1446	0,0148	
6 The 2-nd overheating	275,3	45	2865,3	0,1436		0,047	1
	500		3440,4		7,0323	0,076	
7 The 3-rd overheating	275,9	10	2999,9	0,1196		0,246	
	500		3479,1		7,764	0,354	

Table 5. Steam and water parameters in cycle with 3 re-heating at Pmax=200 gauge atmospheres and Tmax= 500 °C. Theoretical efficiency is 41.7 %, with account of losses on turbine 38.8 %

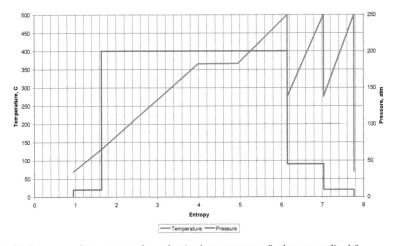

Figure 14. Pressure and temperature dependencies from entropy at final steam quality 1.0.

At variant with common electrical supply network and absence of heat need (for example in the tropics) it is required to take into account turbine cost difference of first and second variants, works financing at NPP building.

In all case it should be noted, that NPP with 39% efficiency is more attractive than other small power NPP variants with efficiency up to 33%. Especially if we take into account many times lower raw uranium requirement and absence of uranium enrichment works for fuel preparation.

On the base of described solutions heavy water gas cooled high power nuclear reactor can be built. Increase of core HWD leads to neutron leakage decrease, which is base neutron loss for nuclear reactor with 80 MW thermal power.

9.2.5. About nuclear safety

Figure 15 shows coolant and moderator ducts scheme in heavy water gas cooled reactor, which prevents power increase at reactivity accident [32].

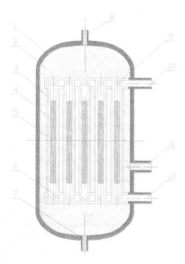

Figure 15. Coolant and moderator ducts scheme in heavy water gas cooled reactor. 1 – reactor vessel, 2 – moderator, 3 – channel casing, 4 – fuel assembly, 5 – channel thermal isolation, 6 – integral coolant collector, 7, 8 – inlet and outlet of moderator, 9 – opening, which connects channel with collector, 10, 12 inlet and outlet of coolant, 11 – nozzle of accident drainage of moderator.

Coolant temperature, which supplies possibility of effective usage of neutron moderation energy, plays here positive role. Moderator temperature increase requires pressure rise in reactor vessel that allows making channel walls thinner and decrease neutron loss in it. Let us take that moderator inlet temperature is 210 °C, outlet temperature 225 °C, flow - 80 kg/s, moderator pressure - 25 atmospheres.

Moderator boiling at unplanned power increase with work pressure and temperature leads to 12 time volume increase of evaporated mass and pressure increase at moderator volume. Core bottom water moves out through nozzle of accident drainage of moderator. Liquid moderator of core upper part is replaced with steam. Preliminary calculations have shown that heavy water deletion from core upper quarter decreases reactivity margin by 2.5 β, that supplies damping of earlier added reactivity.

Damping effect of reactivity accident in this reactor may be not less than in graphite HTGR. Positive feature of this scheme is possibility of work parameters adaptation of safety system by way of specifying work temperature and pressure of moderator, when it starts boiling.

10. About possible scale of nuclear power engineering development

Described actions sufficient using for fission materials reproduction in thermal reactor supplies besides high portion of raw uranium usage (up to 25% in contrast to ~1 % for thermal reactors with enriched fuel) small fuel requirement for core loading. Sum of these effects allows creation of world big nuclear power production industry.

Requirement in raw uranium and thorium of these reactors can be determined by formulae:

$$mU = n * t * \left(m_s * Yu * + m_g * K_i * Ku / 2 \right); \tag{12}$$

$$mTh = n * t * \left(m_s * \left(1 - Yu \right) * + m_g * \left(1 - Ku \right) / 2 \right); \tag{13}$$

where:

n – number of this type built reactor per annum;
t – time of nuclear power engineering development, years;
m_s – mass of raw materials, needed for core loading;
Yu – raw uranium portion in fuel;
m_g – fuel mass needed for year feeding of reactor;
K_i – portion of raw uranium usage in fuel cycle;
Ku – portion of raw uranium in feeding fuel.

Figure 16 shows development variant of power production with even power grow to the level of 8000 GW during 80 years with subsequent power level stabilization.

Work duration of thermal reactors with cheap uranium stocks at this power level and full raw uranium usage is ~2500 years. It is understandable that at such small raw uranium requirements is rational use of other more expensive deposits, where uranium stocks are much more than in cheap deposits.

So, there is significant reserve in world nuclear power production industry.

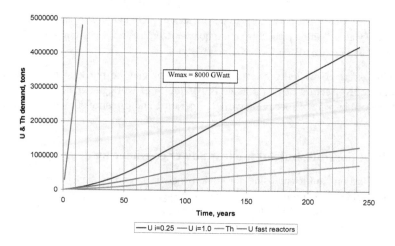

Figure 16. Dependence of uranium and thorium requirements for different reactor types with zero initial power and its even increase up to level of 8000 GW with subsequent power level stabilization.

11. Comparison of possibilities of unauthorized fission materials proliferation at different technologies of fuel cycles

Nonproliferation regime in common case is supplied by IAEA control. With control the possibility of proliferation is limited by value less than detection error. The base of this error is inaccuracy of mass detection at chemical reprocessing of irradiated in core material. Typical value of the error can be in our case ~0.1 % from reprocessed mass of ^{233}U.

For thermal reactor of 1000 MW requirement in ^{233}U is about 10 kg per annum. Accordingly, in ideal case of reprocessing conducting, possible error in ^{233}U is less than 10 g per annum. If inaccuracy of the error is 10 %, than proliferation is not much than 1 g per annum. For minimal warhead in this case is needed more than 1000 years. This term exceeds possible duration not only a terrorist organization but even a terrorist state.

12. Conclusion

Fission materials reproduction possibility in different fuel types in ideal core without neutron losses in construction materials and leakage is shown. Equilibrium concentrations of fission materials in different fuel types are determined.

Features of detailed regime campaign conduction (with fixed fuel location during all campaign) and compact regime (with staged spent fuel replacement with fresh fuel).

Characteristics of loss and reproduction in case of CANDU and its possible modernization variants are examined.

Replacement in zirconium containing materials natural zirconium by isotope ^{90}Zr and natural tin by isotope ^{120}Sn, replacement of 7 fuel rods in fuel assembly with 37 fuel rods by beryllium insert for extra neutron production, change of fuel rods with oxide fuel by metallic fuel is considered as possible modernization directions, which supply high fission materials reproduction.

Compound metallic fuel rod construction, placed in liquid metal heat transferring medium is suggested. Shape and small size of fuel rod ensure decrease of negative effect of swelling.

Possibility of neutron loss decrease in CANDU from 5% to 2.8% in case of isotope modified zirconium and tin and to 1.7% in case of metallic fuel and beryllium insert.

It is suggested that excess neutrons of detailed campaign beginning are used for fission materials reproduction. By the set of characteristics ^{233}U is the best candidate for reproduction.

Portion of raw uranium use increase in open fuel cycle up to 5.3% and full raw uranium usage in closed uranium and thorium fuel cycles is shown.

Conditions of high efficiency obtaining in Rankin cycle with heavy water gas cooled reactor are shown. Scheme of coolant ducts and steam loop of heavy water channel reactor with gaseous coolant, which ensures full use of neutron moderation energy, decrease energy loss for coolant pumping and obtaining high steam quality on the turbine exit. These actions allow to decrease cost for NPP creation, to have efficiency of 43 % taking into account possible losses in core and steam cycle.

World nuclear power production industry creation with power of 8000GW to the end of XXI century on the base of suggested thermal reactors with high fission materials reproduction is shown.

Suggested technologies usage allows increasing world nuclear power industry to the end of XXI century with requirement decrease of natural uranium mining, proliferation danger decrease comparing to fast reactors technology.

Author details

Vladimir M. Kotov

Department of Development and Test of Reactor Devices, Institute of Atomic Energy of NNC RK, Kurchatov, Kazakhstan

Acknowledgement

Author is grateful to R.A. Irkimbekov, S.V. Kotov, A.S. Sergeeva and V.I. Suprunov for help in method discussions and calculations results.

13. References

[1] Adamov E.O., Bolshov L.A., Ganev I.H. et al. Belaya kniga atomnoi energetiki. – Moscow. 2001. – 270 p. (in Russian).

[2] Ponomarev-Stepnoy N.N., Alexeev P.N., Davidenko V.D. et al. Comparison of development directions of nuclear energy in XXI century on the base of material balance. // Atomnaya energiya, 91 (5), 331 (2001) (in Russian).

[3] N.M.Sinev. Economics of nuclear energy. Base of technology and economics of nuclear fuel production Economics of NPP. – Moscow. 1987. Energoatomizdat. - 480 p. (in Russian).

[4] L.V.Matveev, E.M. Center. Uranium-232 and its influence on radiation situation in nuclear fuel cycle. – Moscow. 1983., Energoatomizdat. - 72 p. (in Russian).

[5] Nuclear physics manual. Translation from English, editor L.A. Artcymovich. – Moscow. 1963. State publishing house of physical-mathematical literature. - 632 p. (in Russian).

[6] V.M. Kotov, S.V. Kotov. Fuel nuclides contents change, fission materials reproduction and reactivity margin during thermal reactor campaign with dynamic loading and zone superposition calculation program. // Inv.№ 50, RSE NNC RK, Kurchatov. 2002. (in Russian).

[7] V.M. Kotov, S.V. Kotov, L.N. Tikhomirov Thermal reactor with full use of uranium and thorium fuel creation possibility // Atomnaya energiya, 95 (5), 338-346 (2003) (in Russian).

[8] V.L. Blinkin, V.M. Novikov Molted salt reactors. – Moscow. 1978. Atomizdat. - 111 p. (in Russian).

[9] A.M. Alexeev, Y.M. Bulkin, S.I. Vasiliev at all. High temperature energy technical reactor with solid coolant and radiant heat transfer. // Atomnaya energiya, 56 (1), (1984) (in Russian).

[10] V.M.Kotov, S.V.Kotov, Zh.S.Takibaev, L.N.Tikhomirov. Liquid-salt channel-tipe reactor with dynamic loading and core superposition. / Plasma Devices and Operations. Vol. 13, No. 3, September 2005, 213-221.

[11] B. A. Dementiev. Nuclear power reactors. – Moscow. 1990. Energoatomizdat - 352 p. (in Russian).

[12] F.Ya. Ovchinnikov, V.V. Semenov. Exploitation regimes of water-water power reactors. - Moscow. 1988. Energoatomizdat - 359 p. (in Russian).

[13] The physical theory of neutron chain reactors by Alvin M. Weinberg and Evgene P. Wigner. // The university of Chicago Press. Second edition. 1959.

[14] A.D. Galanin. Introduction to nuclear reactor theory on thermal neutrons. - Moscow. 1984. Energoatomizdat. (in Russian).

[15] V.M. Abramov, V.F. Zelenskiy, M, Kozak et all. First Czechoslovakian atomic power station A-1 with heavy water reactor KS-150 (development and construction). // Atomnaya energiya, 36 (2), 113-124 (1974) (in Russian).

[16] V.A. Gabaraev, A.I. Filin. Development of NPP with reactor BREST – OD-300 with at-station fuel cycle for Beloyarsk NPP site. International scientific-applied conference "Atomic energy and fuel cycles". Moscow – Dimitrovgrad, 1-5 December 2003. (in Russian).

[17] Tin. Popular library of chemical elements. http://n-t.ru/ri/ps/pb050.htm. (in Russian).

[18] A. Radkowsky Applying a thorium-based fuel to non-proliferative commercial light-water reactors. – Nucl. Europe Worldscan, 2000, N 5-6, p.54

[19] MCNP/5: General Monte Carlo N-Particle Transport Code, Version 5, 2003.

[20] V.M. Kotov, R.A. Irkimbekov. Power reactor campaign characteristics calculation. // NNC herald. 3, 118-122 (2011) Kurchatov, (in Russian).

[21] V.M. Kotov, A.S. Dudko, R.A. Irkimbekov. Electro nuclear installation calculation for fission materials production. // NNC herald. 3, (2008) (in Russian).

[22] V.M. Kotov Application of volume neutron source to enhance the use of fertile materials in nuclear power at thermal reactors // Plasma Devices and Operations. 15 (3), 219 (2007).

[23] Gas-cooled power reactors. Directory of Nuclear Reactors. Vol. VII. International Atomic Energy Agency. Vienna, 1968. p.243-302.

[24] A.V. Vasyaev, N.G. Kodochigov, V.M. Rulev et all. Parameters and constructive scheme explanation for energy transfer block with gas turbine cycle in NPP with VTGR. // Atomnaya energiya, 98 (1), 24-36 (2005) (in Russian).

[25] V.M. Kotov, D.I. Zelenskiy. Gas cooled reactor with high efficiency. Intersectoral inter-regional scientific – technical conference " Small power NPP development perspective in regions without central power supply". Moscow, 11-12 November 2010. (in Russian).

[26] V.M. Kotov, G.A. Vityuk, R.A. Irkimbekov., R.A. Mukhametzharova Joint work of heavy water gas cooled reactor with Rankin cycle. // Almaty. International Conference "Nuclear and Radiation Physics", 20-23 September 2011. (in Russian).

[27] Fluent version 6.3.26 User Reference; Fluent, Inc.; 2006.

[28] M.Y. Ivanov Thermal properties of water and steam. Program code Parvo 95. Version 3.3. 2004.// http://fortraner.narod.ru/index.htm (in Russian).

[29] A.A. Rineyskiy Comparison of technical-economical characteristics of NPPs with modern thermal and fast reactors // Atomnaya energiya, 53 (6), 360-367 (1982) (in Russian).

[30] V.M. Kotov Uranium-thorium fuel rod. Innovation patent of Republic of Kazakhstan № 23235 from 20 September 2010. (in Russian).

[31] Udo Zirn, Motonari Haraguchi. Hitachi turbine generator technology for nuclear applications. www.hitachipowersystems.us.

[32] V.M. Kotov Gas cooled reactor with water coolant and way of its control. Innovation patent of Republic of Kazakhstan № 23234 from 20 September 2010. (in Russian).

Chapter 8

On an Analytical Model for the Radioactive Contaminant Release in the Atmosphere from Nuclear Power Plants

Marco Túllio Vilhena, Bardo Bodmann, Umberto Rizza and Daniela Buske

Additional information is available at the end of the chapter

1. Introduction

While the renaissance of nuclear power was motivated by the increasing energy demand and the related climate problem, the recent history of nuclear power, more specifically two disastrous accidents have forced focus on nuclear safety. Although, experience gathered along nuclear reactor developments has sharpened the rules and regulations that lead to the commissioning of latest generation nuclear technology, an issue of crucial concern is the environmental monitoring around nuclear power plants. These measures consider principally the dispersion of radioactive material that either may be released in control actions or in accidents, where in the latter knowledge from simulations guide the planning of emergency actions. In this line the following contribution focuses on the question of radioactive material dispersion after discharge from a nuclear power plant.

The atmosphere is considered the principal vehicle by which radioactive materials that are either released from a nuclear power plant in experimental or eventually in accidental events could be dispersed in the environment and result in radiation exposure of plants, animals and last not least humans. Thus, the evaluation of airborne radioactive material transport in the atmosphere is one of the requirements for monitoring and planning safety measures in the environment around the nuclear power plant. In order to analyse the (possible) consequences of radioactive discharge atmospheric dispersion models are of need, which have to be tuned using specific meteorological parameters and conditions in the considered region. Moreover, they shall be subject to the local orography and supply with realistic information on radiological consequences of routine discharges and potential accidental releases of radioactive substances.

The present work provides a model that allows to implement afore mentioned simulations by the use of a hybrid system. In a first step the local meteorological parameters are determined using the next-generation mesoscale numerical weather prediction system "Weather Research and Forecasting" (WRF). The forcasting system contains a three dimensional data assimilation system and is suitable for applications from the meso- down to the micro-scale. The second step plays the role of simulating the dispersion process in a micro-scale, i.e. in the environment within a radius of several tenth kilometers.

2. On the advection-diffusion approach

The Eulerian approach is widely used in the field of air pollution studies to model the dispersion properties of the Planetary Boundary Layer (PBL). In this context, the diffusion equation that describes the local mean concentrations $\bar{c} = \bar{c}(\mathbf{r}, t)$ at an event point of interest $(\mathbf{r}, t) = (x, y, z, t)$ arising from a any contaminant point source, which may be time dependent, can be written as

$$\partial_t \bar{c} + \mathbf{U}\boldsymbol{\nabla}\bar{c} - \boldsymbol{\nabla}^T \mathbf{K}\boldsymbol{\nabla}\bar{c} = S . \tag{1}$$

Here $\mathbf{U} = (\bar{u}, \bar{v}, \bar{w})^T$ is the vector field of the mean wind velocity, the diagonal matrix $\mathbf{K} = \mathrm{diag}(K_x, K_y, K_z)$ contains the eddy diffusivities and S is a source term, to be determined according to the scenario of interest. In equation (1) we tacitly related the turbulent fluxes $\overline{\mathbf{U}'c'}$ to the gradient of the mean concentration by means of eddy diffusivity (K-theory)

$$\overline{\mathbf{U}'c'} = -\mathbf{K}\boldsymbol{\nabla}\bar{c} \tag{2}$$

The simplicity of the K-theory has led to the widespread use of this theory as mathematical basis for simulating air pollution phenomena. However, the K-closure has its intrinsic limits: it works well when the dimension of dispersed material is much larger than the size of turbulent eddies involved in the diffusion process. Another crucial point is that the down-gradient transport hypothesis is inconsistent with observed features of turbulent diffusion in the upper portion of the mixed layer ([9]). Despite these well known limits, the K-closure is largely used in several atmospheric conditions because it describes the diffusive transport in an Eulerian framework where almost all measurements are easily cast into an Eulerian form, it produces results that agree with experimental data as well as any other more complex model, and it is not computationally expensive as higher order closures usually are.

For a time dependent regime considered in the present work, we assume that the associated advection-diffusion equation adequately describes a dispersion process of radioactive material. From applications of the approach to tracer dispersion data we saw that our analytical approach does not only yield a solution for the three dimensional advection-diffusion equation but predicts tracer concentrations closer to observed values compared to other approaches from the literature, which is also manifest in better statistical coefficients.

Approaches to the advection-diffusion problem are not new in the literature, that are either based on numerical schemes, stochastic simulations or (semi-)analytical methods as shown in a selection of articles ([12, 23, 26, 29, 32]). Note, that in these works all solutions are valid for scenarios with strong restrictions with respect to their specific wind and vertical eddy diffusivity profiles. A more general approach, the ADMM (Advection Diffusion Multilayer Method) approach solves the two-dimensional advection-diffusion equation with variable wind profile and eddy diffusivity coefficient ([21]). The main idea here relies on the discretisation of the atmospheric boundary layer in a multi-shell domain, assuming in each layer that eddy diffusivity and wind profile take averaged values. The resulting advection-diffusion equation in each layer is then solved by the Laplace Transform technique. The GIADMT method (Generalized Integral Advection Diffusion Multilayer Technique) ([7]) is a dimensional extension to the previous work, but again assuming the stepwise approximation for the eddy diffusivity coefficient and wind profile. To generalize, a general two-dimensional solution was presented by ([22]). The solving methodology was the Generalized Integral Laplace Transform Technique (GILTT) that is an analytical series solution including the solution of an associate Sturm-Liouville problem, expansion of the pollutant concentration in a series in terms of the attained eigenfunction, replacement of this expansion in the advection-diffusion equation and, finally, taking moments. This procedure leads to a set of differential ordinary equations that is solved analytically by Laplace transform technique. In this work we improve further the solutions of the afore mentioned articles and report on a general analytical solution for the advection-diffusion problem, assuming that eddy diffusivity and wind profiles are arbitrary functions having a continuous dependence on the vertical and longitudinal spatial variables.

Equation (1) is considered valid in the domain $(x, y, z) \in \Gamma$ bounded by $0 < x < L_x, 0 < y < L_y$ (with L_x and L_y sufficiently large), $0 < z < h$ (here h is the boundary layer height) and subject to the following boundary and initial conditions,

$$\mathbf{K}\nabla \bar{c}|_{(0,0,0)} = \mathbf{K}\nabla \bar{c}|_{(L_x,L_y,h)} = 0 \tag{3}$$

$$\bar{c}(x, y, z, 0) = 0 . \tag{4}$$

Instead of specifying the source term as an inhomogeneity of the partial differential equation, we consider a point source located at an edge of the domain, so that the source position $\mathbf{r}_S = (0, y_0, H_S)$ is located at the boundary of the domain $\mathbf{r}_S \in \delta\Gamma$. Note, that in cases where the source is located in the domain, one still may divide the whole domain in sub-domains, where the source lies on the boundary of the sub-domains which can be solved for each sub-domain separately. Moreover, a set of different sources may be implemented as a superposition of independent problems. Since the source term location is on the boundary, in the domain this term is zero everywhere ($S(\mathbf{r}) \equiv 0$ for $\mathbf{r} \in \Gamma \backslash \delta\Gamma$), so that the source influence may be cast in form of a condition rather than a source term of the equation. The source condition for a time dependent contamination reads then

$$S = \oint \omega_S \, d\Sigma \tag{5}$$

where ω_S represents a flux across a closed surface that includes the source and is proportional to the source strength. Instead of considering an explicit source term, we implement the solution as a superposition of an infinite number of solutions with instantaneous source represented in an initial condition. The solution for a time dependent source assumes the following form

$$\bar{c}(t, x, y, z) = \int_0^t \dot{\bar{c}}(t - \tau, x, y, z) \, d\tau \tag{6}$$

with instantaneous initial condition

$$\dot{\bar{c}}(0, x, y, z) = \dot{\bar{c}}_0 = \lim_{\oint \hat{\Sigma} \, d\Sigma \to 0} \oint \omega_S \, d\Sigma = Q\delta(x)\delta(y - y_0)\delta(z - H_S) \tag{7}$$

where Q is the emission rate, H_S the height of the source, $\delta(x)$ represents the Cartesian Dirac delta functional and $\hat{\boldsymbol{\Sigma}}$ is a unit vector.

3. A closed form solution

In this section we first introduce the general formalism to solve a general problem and subsequently reduce the problem to a more specific one, that is solved and compared to experimental findings.

3.1. The general procedure

In order to solve the problem (1) we reduce the dimensionality by one and thus cast the problem into a form already solved in reference [22]. To this end we apply the integral transform technique in the y variable, and expand the pollutant concentration as

$$\bar{c}(x,y,z,t) = \mathbf{R}^T(x,z,t)\mathbf{Y}(y), \tag{8}$$

where $\mathbf{R} = (R_1, R_2, \ldots)^T$ and $\mathbf{Y} = (Y_1, Y_2, \ldots)^T$ is a vector in the space of orthogonal eigenfunctions, given by $Y_m(y) = \cos(\lambda_m y)$ with eigenvalues $\lambda_m = m\frac{\pi}{L_y}$ for $m = 0, 1, 2, \ldots$. For convenience we introduce some shorthand notations, $\nabla_2 = (\partial_x, 0, \partial_y)^T$ and $\hat{\partial}_y = (0, \partial_y, 0)^T$, so that equation (1) reads now,

$$(\partial_t \mathbf{R}^T)\mathbf{Y} + \bar{\mathbf{U}}\left(\nabla_2 \mathbf{R}^T\mathbf{Y} + \mathbf{R}^T\hat{\partial}_y\mathbf{Y}\right) = \left(\nabla^T\mathbf{K} + (\mathbf{K}\nabla)^T\right)\left(\nabla_2\mathbf{R}^T\mathbf{Y} + \mathbf{R}^T\hat{\partial}_y\mathbf{Y}\right)$$
$$= \left(\nabla_2^T\mathbf{K} + (\mathbf{K}\nabla_2)^T\right)(\nabla_2\mathbf{R}^T\mathbf{Y}) + \left(\hat{\partial}_y^T\mathbf{K} + (\mathbf{K}\hat{\partial}_y)^T\right)(\mathbf{R}^T\hat{\partial}_y\mathbf{Y}). \tag{9}$$

Upon application of the integral operator

$$\int_0^{L_y} dy \mathbf{Y}[\mathbf{F}] = \int_0^{L_y} \mathbf{F}^T \wedge \mathbf{Y}\, dy \tag{10}$$

here $\mathbf{F}$ is an arbitrary function and $\wedge$ signifies the dyadic product operator, and making use of orthogonality renders equation (9) a matrix equation. The appearing integral terms are

$$\mathbf{B}_0 = \int_0^{L_y} dy \mathbf{Y}[\mathbf{Y}] = \int_0^{L_y} \mathbf{Y}^T \wedge \mathbf{Y}\, dy,$$

$$\mathbf{Z} = \int_0^{L_y} dy \mathbf{Y}[\hat{\partial}_y \mathbf{Y}] = \int_0^{L_y} \hat{\partial}_y \mathbf{Y}^T \wedge \mathbf{Y}\, dy,$$

$$\Omega_1 = \int_0^{L_y} dy \mathbf{Y}[(\nabla_2^T\mathbf{K})(\nabla_2\mathbf{R}^T\mathbf{Y})] = \int_0^{L_y} \left((\nabla_2^T\mathbf{K})(\nabla_2\mathbf{R}^T\mathbf{Y})\right)^T \wedge \mathbf{Y}\, dy, \tag{11}$$

$$\Omega_2 = \int_0^{L_y} dy \mathbf{Y}[(\mathbf{K}\nabla_2)^T(\nabla_2\mathbf{R}^T\mathbf{Y})] = \int_0^{L_y} \left((\mathbf{K}\nabla_2)^T(\nabla_2\mathbf{R}^T\mathbf{Y})\right) \wedge \mathbf{Y}\, dy,$$

$$\mathbf{T}_1 = \int_0^{L_y} dy \mathbf{Y}[((\hat{\partial}_y^T\mathbf{K})(\hat{\partial}_y\mathbf{Y})] = \int_0^{L_y} \left(((\hat{\partial}_y^T\mathbf{K})(\hat{\partial}_y\mathbf{Y}))\right)^T \wedge \mathbf{Y}\, dy,$$

$$\mathbf{T}_2 = \int_0^{L_y} dy \mathbf{Y}[(\mathbf{K}\hat{\partial}_y)^T(\hat{\partial}_y\mathbf{Y})] = \int_0^{L_y} \left((\mathbf{K}\hat{\partial}_y)^T(\hat{\partial}_y\mathbf{Y})\right)^T \wedge \mathbf{Y}\, dy.$$

Here, $\mathbf{B}_0 = \frac{L_y}{2}\mathbf{I}$, where $\mathbf{I}$ is the identity, the elements $(\mathbf{Z})_{mn} = \frac{2}{1-n^2/m^2}\delta_{1,j}$ with $\delta_{i,j}$ the Kronecker symbol and $j = (m+n)\mathrm{mod2}$ is the remainder of an integer division (i.e. is one for $m+n$ odd and zero else). Note, that the integrals Ω_i and $\mathbf{T}_i$ depend on the specific form of the eddy diffusivity $\mathbf{K}$. The integrals (11) are general, but for practical purposes and for application to a case study we truncate the eigenfunction space and consider M components in $\mathbf{R}$ and $\mathbf{Y}$ only, though continue using the general nomenclature that remains valid. The obtained matrix equation determines now together with initial and boundary condition uniquely the components R_i for $i = 1, \ldots, M$ following the procedure introduced in reference [22]:

$$(\partial_t \mathbf{R}^T)\mathbf{B} + \bar{\mathbf{U}}\left(\nabla_2 \mathbf{R}^T \mathbf{B} + \mathbf{R}^T \mathbf{Z}\right) = \Omega_1(\mathbf{R}) + \Omega_2(\mathbf{R}) + \mathbf{R}^T(\mathbf{T}_1 + \mathbf{T}_2) \qquad (12)$$

3.2. A specific case for application

In order to discuss a specific case we introduce a convention and consider the average wind velocity $\bar{\mathbf{U}} = (\bar{u}, 0, 0)^T$ aligned with the x-axis. Since the variation of the average wind velocity is slow compared to the time intervals for which the meteorological data are extracted from WRF, we superimpose the solution after rotation in the $x - y$-plane in order to transform every instantaneous solution into the same coordinate frame, i.e. the coordinate frame for $t = 0$. By comparison of physically meaningful cases, one finds for the operator norm $||\partial_x K_x \partial_x|| <<|\bar{u}|$, which can be understood intuitively because eddy diffusion is observable predominantly perpendicular to the mean wind propagation. As a consequence we neglect the terms with K_x and $\partial_x K_x$.

The principal aspect of interest in pollution dispersion is the vertical concentration profile, that responds strongly to the atmospheric boundary layer stratification, so that the simplified eddy diffusivity $\mathbf{K} \to \mathbf{K}_1 = \mathrm{diag}(0, K_y, K_z)$ depends in leading order approximation only on the vertical coordinate $\mathbf{K}_1 = \mathbf{K}_1(z)$. For this specific case the integrals Ω_i reduce to

$$\Omega_1 \to (\partial_z K_z)(\partial_z \mathbf{R}^T)\mathbf{B} ,$$
$$\Omega_2 \to K_z(\partial_z^2 \mathbf{R}^T)\mathbf{B} , \qquad (13)$$
$$\mathbf{T}_1 \to \mathbf{0} ,$$
$$\mathbf{T}_2 \to -K_y \Lambda \mathbf{B} , \qquad (14)$$

where $\Lambda = \mathrm{diag}(\lambda_1^2, \lambda_2^2, \ldots)$. The simplified equation system to be solved is then,

$$\partial_t \mathbf{R}^T \mathbf{B} + \bar{u}\partial_x \mathbf{R}^T \mathbf{B} = (\partial_z K_z)\partial_z \mathbf{R}^T \mathbf{B} + K_z\partial_z^2 \mathbf{R}^T \mathbf{B} - K_y \mathbf{R}^T \Lambda \mathbf{B} \qquad (15)$$

which is equivalent to the problem

$$\partial_t \mathbf{R} + \bar{u}\partial_x \mathbf{R} = (\partial_z K_z)\partial_z \mathbf{R} + K_z\partial_z^2 \mathbf{R} - K_y \Lambda \mathbf{R} \qquad (16)$$

by virtue of $\mathbf{B}$ being a diagonal matrix.

The specific form of the eddy diffusivity determines now whether the problem is a linear or non-linear one. In the linear case the $\mathbf{K}$ is assumed to be independent of $\bar{c}$, whereas in more realistic cases, even if stationary, $\mathbf{K}$ may depend on the contaminant concentration and thus renders the problem non-linear. However, until now now specific law is known that links the eddy diffusivity to the concentration so that we hide this dependence using a phenomenologically motivated expression for $\mathbf{K}$ which leaves us with a partial differential equation system in linear form, although the original phenomenon is non-linear. In the example below we demonstrate the closed form procedure for a problem with explicit time dependence, which is novel in the literature.

The solution is generated making use of the decomposition method ([1–3]) which was originally proposed to solve non-linear partial differential equations, followed by the Laplace transform that renders the problem a pseudo-stationary one. Further we rewrite the vertical diffusivity as a time average term $\bar{K}_z(z)$ plus a term representing the variations $\kappa_z(z,t)$ around the average for the time interval of the measurement $K_z(x,z,t) = \bar{K}_z(z) + \kappa_z(z,t)$ and use the asymptotic form of K_y, which is then explored to set-up the structure of the equation that defines the recursive decomposition scheme:

$$\partial_t \mathbf{R} + \bar{u}\partial_x \mathbf{R} - \partial_z\left(\bar{K}_z\partial_z\mathbf{R}\right) + K_y\Lambda\mathbf{R} = \partial_z\left(\kappa_z\partial_z\mathbf{R}\right) \tag{17}$$

The function $\mathbf{R} = \sum_j \mathbf{R}_j = \mathbf{1}^T\mathbf{R}^{(c)}$ is now decomposed into contributions to be determined by recursion. For convenience we introduced the one-vector $\mathbf{1} = (1,1,\dots)^T$ and inflate the vector $\mathbf{R}$ to a vector with each element being itself a vector $\mathbf{R}_j$. Upon inserting the expansion in equation (17) one may regroup terms that obey the recursive equations and starts with the time averaged solution for K_z:

$$\partial_t \mathbf{R}_0 + \bar{u}\partial_x \mathbf{R}_0 - \partial_z\left(\bar{K}_z\partial_z\mathbf{R}_0\right) + K_y\Lambda\mathbf{R}_0 = 0 \tag{18}$$

The extension to the closed form recursion is then given by

$$\partial_t \mathbf{R}_j + \bar{u}\partial_x \mathbf{R}_j - \partial_z\left(\bar{K}_z\partial_z\mathbf{R}_j\right) + K_y\Lambda\mathbf{R}_j = \partial_z\left(\kappa_z\partial_z\mathbf{R}_{j-1}\right). \tag{19}$$

From the construction of the recursion equation system it is evident that other schemes are possible. The specific choice made here allows us to solve the recursion initialisation using the procedure described in reference [22], where a stationary $\mathbf{K}$ was assumed. For this reason the time dependence enters as a known source term from the first recursion step on.

3.3. Recursion initialisation

The boundary conditions are now used to uniquely determine the solution. In our scheme the initialisation solution that contains $\mathbf{R}_0$ satisfies the boundary conditions (equations (3)) while the remaining equations satisfy homogeneous boundary conditions. Once the set of problems (19) is solved by the GILTT method, the solution of problem (1) is well determined. It is important to consider that we may control the accuracy of the results by a proper choice of the number of terms in the solution series.

In reference [22] a two dimensional problem with advection in the x direction in stationary regime was solved which has the same formal structure than (19) except for the time dependence. Upon rendering the recursion scheme in a pseudo-stationary form problem and thus matching the recursive structure of [22], we apply the Laplace Transform in the t variable, $(t \to r)$ obtaining the following pseudo-steady-state problem:

$$r\tilde{\mathbf{R}}_0 + \bar{u}\partial_x\tilde{\mathbf{R}}_0 = \partial_z\left(K_z\partial_z\tilde{\mathbf{R}}_0\right) - \Lambda K_y\tilde{\mathbf{R}}_0 \tag{20}$$

The x and z dependence may be separated using the same reasoning as already introduced in (8). To this end we pose the solution of problem (20) in the form:

$$\tilde{R}_0 = PC \tag{21}$$

where $C = (\zeta_1(z), \zeta_2(z), \dots)^T$ are a set of orthogonal eigenfunctions, given by $\zeta_i(z) = \cos(\gamma_l z)$, and $\gamma_i = i\pi/h$ (for $i = 0, 1, 2, \dots$) are the set of eigenvalues.

Replacing equation (21) in equation (20) and using the afore introduced projector (10) now for the z dependent degrees of freedom $\int_0^h dz C[F] = \int_0^h F^T \wedge C\, dz$ yields a first order differential equation system:

$$\partial_x P + HP = 0 , \tag{22}$$

where $P = P(x, r)$ and $H = B_1^{-1} B_2$. The entries of matrices B_1 and B_2 are

$$(B_1)_{i,j} = -\int_0^h \bar{u}\zeta_i(z)\zeta_j(z)\, dz$$

$$(B_2)_{i,j} = \int_0^h \partial_z K_z \partial_z \zeta_i(z)\zeta_j(z)\, dz - \gamma_i^2 \int_0^h K_z \zeta_i(z)\zeta_j(z)\, dz$$

$$-r\int_0^h \zeta_i(z)\zeta_j(z)\, dz - \lambda_i^2 K_y \int_0^h \zeta_i(z)\zeta_j(z)\, dz .$$

A similar procedure leads to the source condition for (22):

$$P(0, r) = Q B_1^{-1} \int dz C[\delta(z - H_S)] \int dy Y[\delta(y - y_0)] = Q B_1^{-1} (C(H_S) \wedge 1)(1 \wedge Y(y_0)) \tag{23}$$

Following the reasoning of [22] we solve (22) applying Laplace transform and diagonalisation of the matrix $H = XDX^{-1}$ which results in

$$\tilde{P}(s, r) = X(sI + D)^{-1} X^{-1} P(0, r) \tag{24}$$

where $\tilde{P}(s, r)$ denotes the Laplace Transform of $P(x, r)$. Here $X^{(-1)}$ is the (inverse) matrix of the eigenvectors of matrix $B_1^{-1} B_2$ with diagonal eigenvalue matrix D and the entries of matrix $(sI + D)_{ii} = s + d_i$. After performing the Laplace transform inversion of equation (24), we come out with

$$P(x, r) = XG(x, r) X^{-1} \Omega , \tag{25}$$

where $G(x, r)$ is the diagonal matrix with components $(G)_{ii} = e^{-d_i x}$. Further the still unknown arbitrary constant matrix is given by $\Omega = X^{-1} P(0, r)$.

The analytical time dependence for the recursion initialisation (20) is obtained upon applying the inverse Laplace transform definition

$$R_0(x, z, t) = \frac{1}{2\pi i} \int_{\gamma - i\infty}^{\gamma + i\infty} P(x, r) C(z) e^{rt}\, dr . \tag{26}$$

To overcome the drawback of evaluating the line integral appearing in the above solution, we perform the calculation of this integral by the Gaussian quadrature scheme, which is exact if the integrand is a polynomial of degree $2M - 1$ in the $\frac{1}{r}$ variable

$$R_0(x, z, t) = \frac{1}{t} a^T \left(p R_0\left(x, z, \frac{p}{t}\right) \right) , \tag{27}$$

where **a** and **p** are respectively vectors with the weights and roots of the Gaussian quadrature scheme ([27]), and the argument $(x, z, \frac{p}{r})$ signifies the k-th component of **p** in the k-th row of $p R_0$. Note, k is a component from contraction with **a**.

4. Experimental data and turbulent parameterisation

For model validation we chose a controlled release of radioactive material performed in 1985 at the Itaorna Beach, close to the nuclear reactor site Angra dos Reis in the Rio de Janeiro state, Brazil. Details of the dispersion experiment is described elsewhere ([5]). The experiment consisted in the controlled releases of radioactive tritiated water vapour from the meteorological tower at $100m$ height during five days (28 November to 4 December 1984). During the whole experiment, four meteorological towers collected the relevant meteorological data. Wind speed and direction were measured at three levels ($10m$, $60m$ and $100m$) together with the temperature gradients between $10m$ and $100m$. Some additional data of relative humidity were available in some of the sampling sites, and were used to calculate the concentration of radioactive tritiated water in the air (after measuring the radioactivity of the collected samples). All relevant details, as well as the synoptic meteorological conditions during the dispersion campaign are described in ref. [5]. The data from experiments 2 and 3 were used to obtain the numerical results and are presented in table 1.

Exp	Period	$U(m/s)$	$h(m)$	$u_*(m/s)$	$L(m)$	$w_*(m/s)$	$Q(MBq/s)$
2	3	2.2	1134	0.4	-951	0.6	25.3
3	3	2.6	1367	0.5	-1147	0.7	20.5

Table 1. Micro-meteorological parameters and emission rate for experiments 2 and 3 at third period.

The micro-meteorological parameters shown in table 1 are calculated from equations obtained in the literature. The roughness length utilized was $1m$ and the Monin-Obukhov length for convective conditions can be written as $L = -h/k \left(u_*/w_* \right)^3$ ([35]), where k is the von Karman constant ($k = 0.4$), w_* is the convective velocity scale with wind speed U, $u_* = kU/\ln(z_r/z_0)$ is the friction velocity, where U is the wind velocity at the reference height $z_r = 10m$, and $h = 0.3u_*/f_c$ is the height of the boundary layer with the Coriolis coefficient $f_c = 10^{-4}$.

In the atmospheric diffusion problems the choice of a turbulent parameterisation represents a fundamental aspect for contaminant dispersion modelling. From the physical point of view a turbulence parameterisation an approximation for the natural phenomenon, where details are hidden in the parameters used, that have to be adjusted in order to reproduce experimental findings. The reliability of each model strongly depends on the way the turbulent parameters are calculated and related to the current understanding of the planetary boundary layer. In terms of the convective scaling parameters the vertical and lateral eddy diffusivities can be formulated as follows ([11]):

$$K_z = 0.22 w_* h \left(\frac{z}{h} \right)^{\frac{1}{3}} \left(1 - \frac{z}{h} \right)^{\frac{1}{3}} \left(1 - e^{\frac{4z}{h}} - 0.0003 e^{\frac{8z}{h}} \right) \qquad (28)$$

$$K_y = \frac{\sqrt{\pi}\sigma_v}{16(f_m)_v q_v} \qquad \text{with} \qquad \sigma_v^2 = \frac{0.98c_v}{(f_m)_v^{\frac{2}{3}}} \left(\frac{\psi_\epsilon}{q_v}\right)^{\frac{2}{3}} \left(\frac{z}{h}\right)^{\frac{2}{3}} w_*^2$$

$$q_v = 4.16\frac{z}{h}, \qquad \psi_\epsilon^{\frac{1}{3}} = \left(\left(1-\frac{z}{h}\right)^2 \left(-\frac{z}{L}\right)^{-\frac{2}{3}} + 0.75\right)^{\frac{1}{2}} \qquad \text{and} \qquad (f_m)_v = 0.16 \qquad (29)$$

where σ_v is the standard deviation of the longitudinal turbulent velocity component, q_v is the stability function, ψ_ϵ is the dimensionless molecular dissipation rate and $(f_m)_v$ is the transverse wave peak.

The wind speed profile can be described by a power law $u_z/u_1 = (z/z_1)^n$ ([25]), where u_z and u_1 are the horizontal mean wind speeds at heights z and z_1 and n is an exponent that is related to the intensity of turbulence ([16]).

Thus, in this study we introduce the vertical and lateral eddy diffusivities (eq. (35) and eq. (29)) and the power law wind profile in the 3D-GILTT model (eq. (16) or equivalently eq. (20)) to calculate the ground-level concentration of emissions released from an elevated continuous source point in an unstable/neutral atmospheric boundary layer.

The validation of the 3D-GILTT model predictions against experimental data from the Angra site together with a two dimensional model (GILTTG) are shown in table 2. While the present approach (3D-GILTT) is based on a genuine three dimensional description an earlier analytical approach (GILTTG) uses a Gaussian assumption for the horizontal transverse direction ([22]). Figure 1 shows the comparison of predicted concentrations against observed ones for the three dimensional approach, which reproduces acceptably the observed concentrations, although this simulation did not make use of the terrain's realistic complexity.

In the further we use the standard statistical indices in order to compare the quality of the two approaches. Note, that we present the two analytical model approaches, since the earlier

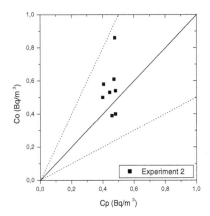
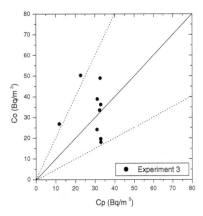

Figure 1. Observed and predicted scatter diagram of ground-level concentrations using the 3D-GILTT approach for the experiment; dotted lines indicate a factor of two.

Exp.	Period	Distance (m)	Observed (Bq/m^3)	Predictions (Bq/m^3)	
				GILTTG	3D-GILTT
2	3	610	0.58	0.20	0.40
2	3	600	0.50	0.19	0.40
2	3	700	0.53	0.29	0.44
2	3	815	0.61	0.38	0.47
2	3	970	0.54	0.47	0.48
2	3	1070	0.86	0.51	0.48
2	3	750	0.39	0.33	0.46
2	3	935	0.40	0.45	0.48
3	3	705	38.89	47.18	31.13
3	3	700	24.09	46.53	31.02
3	3	815	48.95	59.98	32.66
3	3	970	36.22	73.03	32.95
3	3	1070	33.50	78.65	32.44
3	3	500	50.26	17.74	22.58
3	3	375	26.86	2.57	11.67
3	3	960	19.61	72.35	32.97
3	3	915	18.02	69.04	33.03

Table 2. Concentrations of nine runs with various positions of the Angra dos Reis experiment and model prediction by the approaches GILTTG and 3D-GILTT.

one was found to be acceptable in comparison to other approaches found in the literature and both give a solution in closed form. The standard statistical indices are NMSE, the normalized mean square error; COR, the correlation coefficient; FA2 and FA5, the fraction of data (in %) in the cones determined by a factor of two and five, respectively; FB, the fractional bias and FS, the fractional standard deviation. The subscripts o and p refer to observed and predicted quantities, respectively, and $\bar{C}$ indicates the averaged values. Table 3 presents the results of the statistical indices used to evaluate the model performance ([14]) and further compare our model to the GILTTG approach. The statistical index FB indicates weather the predicted quantities (C_p) under- or overestimates the observed ones (C_o). The statistical index NMSE represents the quadratic error of the predicted quantities in relation to the observed ones. Best results are indicated by values compatible with zero for NMSE, FB and FS, and compatible with unity for COR, FA2 and FA5. The statistical indices point out that a reasonable agreement is obtained between experimental data and the 3D-GILTT model.

In order to validate the two models we fit the predicted versus observed values by a linear regression (see figure 2), where the closer their intersect to the origin and the closer the slope is to unity the better is the approach. The GILTTG approach results in $\bar{C}_p = 1.16\bar{C}_o + 7.01$

Statistical Indices	GILTTG	3D-GILTT
NMSE $= \overline{(C_o - C_p)^2} / \overline{C_p}\, \overline{C_o}$	1.34	0.38
COR $= \overline{(C_o - \overline{C_o})(C_p - \overline{C_p})} / \sigma_o \sigma_p$	0.67	0.83
FA2 $= 0.5 \leq (C_p / C_o) \leq 2$	0.53	0.88
FA5 $= 0.2 \leq (C_p / C_o) \leq 5$	0.96	1.00
FB $= \overline{C_o} - \overline{C_p} / 0.5(\overline{C_o} + \overline{C_p})$	−0.44	0.13
FS $= (\sigma_o - \sigma_p) / 0.5(\sigma_o + \sigma_p)$	−0.54	0.18

Table 3. Statistical comparisons between GILTTG and 3D-GILTT results.

with $R^2 = 0.67$ and $\kappa = 0.43$, whereas the 3D-GILTT obeys the result $\bar{C}_p = 0.69\bar{C}_o + 3.26$ with $R^2 = 0.83$ and $\kappa = 0.36$. In order to perform a model validation we introduced an index $\kappa = \sqrt{(a - 1)^2 + (b/\bar{C}_o)^2}$ with $\bar{C}_o = \frac{1}{n} \sum_{i=1}^{n} C_{oi}$, which if identical zero indicates a perfect match between the model and the experimental findings. Here a is the slope, b the intersection, C_{oi} of the experimental data and $\bar{C}_o$ its arithmetic mean. Since the experiment is of stochastic character whereas the stochastic properties are hidden in the model parameters, considerable fluctuations are present. Nevertheless, by comparison one observes that the present approach yields the better description of the data.

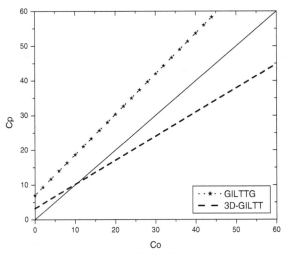

Figure 2. Linear regression for the GILTTG and 3D-GILTT. The bisector was added as an eye guide.

5. Meso-scale simulation for K-closure

The consistency of the K-approach strongly depends on the way the eddy diffusivity is determined on the basis of the turbulence structure of the PBL and on the model ability to reproduce experimental diffusion data. Keeping the K-theory limitations in mind many efforts have been made to develop turbulent parametrisations for practical applications in air pollution modelling which reveals the essential features of turbulent diffusion, but which as far as possible preserves the simplicity and flexibility of the K-theory formulation. The aim of this step is to elaborate parametrisations for the eddy diffusivity coefficients in the PBL based on the micro-meteorological parameters that were extracted from mesoscale WRF simulations. The WRF model is based on the Taylor's statistical theory and a model for Eulerian spectra ([11, 24]). The main idea of the proposed spectral model relies on considering the turbulent spectra as a superposition of a buoyant produced part (with a convective peak wavelength) and a shear produced part (with a mechanical peak wavelength). By such a model, the plume spreading rate is directly connected with the spectral distribution of eddies in the PBL, that is with the energy containing eddies of the turbulence.

The WRF Simulator is a meso-scale numerical weather prediction system that features multiple dynamical cores and a 3-dimensional variational data assimilation system. The simulator offers multiple physics options that can be combined in various ways. Since this study focusses on the implementation of an interface with a model for the PBL, orography related features of WRF were of importance, more specifically the Land-Surface and PBL physics options were chosen for the present study. In WRF, when a PBL scheme is activated, a specific vertical diffusion is de-activated with the assumption that the PBL scheme will handle this process. The Mellor-Yamada-Janjic PBL scheme derives the eddy diffusivities coefficients and the boundary layer height from the estimations of the Turbulent Kinetic Energy (TKE) through the full range of atmospheric turbulent regimes ([19]).

Two grids were used for the WRF meso-scale simulation. The outer grid has an extension of the order of half the earth radius so that a significant part of the large scale geological domain of interest is included. The inner grid is centred at the point of interest, i.e. the centre of the power plant where typically the nuclear reactor is located. The simulation may in principal contain a sequence of days or even months. The micro-meteorological data are extracted at the centre point of the inner WRF grid. The spectral model needs these quantities to calculate the eddy diffusivity coefficients.

On the basis of Taylor's theory, Taylor proposed that under the hypothesis of homogeneous turbulence, the eddy diffusivities may be expressed as

$$K_\alpha = \frac{d}{dt}\left(\frac{\sigma_\alpha^2}{2}\right) = \frac{\sigma_i^2 \beta_i}{2\pi} \int_0^\infty F_i^E(n) \frac{\sin(2\pi n t \beta_i^{-1})}{n} dn, \qquad (30)$$

where $\alpha = (x, y, z)$ and $i = u, v, w$, $F_i^E(n)$ is the value of the Eulerian spectrum of energy normalized by the Eulerian velocity variance, and σ_i^2 corresponds to the Eulerian variance of the turbulent wind field. Following [33], $\beta_i = \left(\frac{\pi U^2}{16\sigma_i^2}\right)^{\frac{1}{2}}$. For large diffusion travel times ($t \to \infty$), the filter function in the integral of eqn. (30) selects $F_i^E(n)$ at the origin of the frequency space, such that the rate of dispersion becomes independent of the travel time from the source and can be expressed as a function of local properties of turbulence,

$$K_\alpha = \frac{\sigma_i^2 \beta_i F_i^E(0)}{4} \qquad (31)$$

where $F_i^E(0)$ is the value of the normalised Eulerian energy spectrum at $n = 0$. In this way the eddy diffusivity is directly associated to the energy-containing eddies which are the principal contribution to turbulent transport. In order to use eqn. (31) we have to find an analytical form for the dimensionless Eulerian spectrum. We assume here that the spectral distribution of turbulent kinetic energy is a superposition of buoyancy and shear components. Such a TKE model may be evaluated as a good approximation for a real PBL, where turbulent production is due to both mechanisms ([15, 20]). In these conditions we may write the Eulerian dimensional spectrum as $S_i^E(n) = S_{ib}(n) + S_{is}(n)$, where the subscripts b and s stand for buoyancy and shear, respectively.

An analytical form for the dimensional spectra in convective turbulence has been reported in [11]

$$S_{ib}(n) = \frac{0.98c_i \left(\frac{nz}{\bar{u}}\right)}{n \left(f_{mi}^*\right)^{\frac{5}{3}} \left(1+1.5\frac{\frac{nz}{\bar{u}}}{f_{mi}^*}\right)} \Psi_{\epsilon b}^{\frac{2}{3}} \left(\frac{z}{z_i}\right)^{\frac{2}{3}} w_*^2 , \tag{32}$$

while for mechanical turbulence ([10])

$$S_{is}(n) = \frac{1.5c_i \left(\frac{nz}{\bar{u}}\right)}{n \left(f_{mi}\right)^{\frac{5}{3}} \left(1+1.5\frac{\frac{nz}{\bar{u}}}{f_{mi}}\right)} \Phi_{\epsilon s}^{\frac{2}{3}} u_*^2 \tag{33}$$

where $\Psi_{\epsilon b} = \frac{\epsilon_b h}{w_*^3}$ and $\Phi_{\epsilon s} = \frac{\epsilon \kappa z}{u_*^3}$ are the dimensional dissipation rate functions, ϵ_b and ϵ_s are the convective and mechanical rate of tke dissipation, f_{mi}^* is the normalized frequency of the spectral peaks regardless of stratification and f_{mi} is the reduced frequency with the mean wind speed $\bar{u}$ in the mixing layer.

The dimensionless spectrum $F_i^E(n)$ in eqn. (31) is obtained by normalizing the dimensional spectra with the total variance, $\sigma_i^2 = \int_0^\infty S_i^E(n) dn$, that is

$$F_i^E(n) = \frac{S_i^E}{\sigma_i^2} = \frac{S_{ib}^E(n) + S_{is}^E(n)}{\sigma_i^2} . \tag{34}$$

The total wind velocity variance is obtained by the sum of mechanical and convective variances $\sigma_i^2 = \int_0^\infty (S_{ib}^E(n) + S_{is}^E(n)) \, dn = \sigma_{ib}^2 + \sigma_{is}^2$. Making use of eqns. (30), (32), (33) and eqn. (34) one ends up with $K_\alpha = \frac{\beta_i}{4} \left(S_{ib}^E(0) + S_{is}^E(0)\right)$, that for the w-component becomes

$$K_z = \frac{\beta_i}{4} \left(\frac{0.98c_w \left(\frac{z}{\bar{u}}\right)}{n(f_{mw}^*)^{\frac{5}{3}}} \Psi_{\epsilon b}^{\frac{2}{3}} \left(\frac{z}{z_i}\right)^{\frac{2}{3}} w_*^2 + \frac{1.5c_w \left(\frac{z}{\bar{u}}\right)}{n(f_{mw})^{\frac{5}{3}}} \Phi_{\epsilon s}^{\frac{2}{3}} u_*^2 \right) \tag{35}$$

6. Application to the Fukushima-Daiichi accident

In order to illustrate the suitability of the discussed formulation to simulate contaminant dispersion in the atmospheric boundary layer, we evaluate the performance of the new solution and simulate radioactive substance dispersion around the Fukushima-Daiichi power plant.

At the 11th of march, 2011 the Fukushima-Daiichi nuclear power plant accident (coordinates in latitude, longitude: 37^0 25' 17" N, 141^0 1' 57" E) caused considerable radiation leakage into the atmosphere and into the sea. The radioactive pollution of the environment and sea was caused principally by the direct release of contaminated water from the power station. To a lesser extent atmospheric release of the radio-nuclide from the atmospheric plume are carried by the winds over the sea during and after the accident sequence. Shorter–lived radioactive elements, such as Iodine-131 were detectable for a few months (half-live of approximately 80 days). Others, such as Ruthenium-106 and Caesium-134 will still persist in the environment for several years (Caesium-137 has a half-life of approximately 30 years).

In the following we show the results for a sequence of four days from the 12th to the 15th of march. Figure 3 shows some meso-scale meteorological information, that was obtained from WRF. The first plot in fig. 3 corresponds to the situation three hours after the beginning of

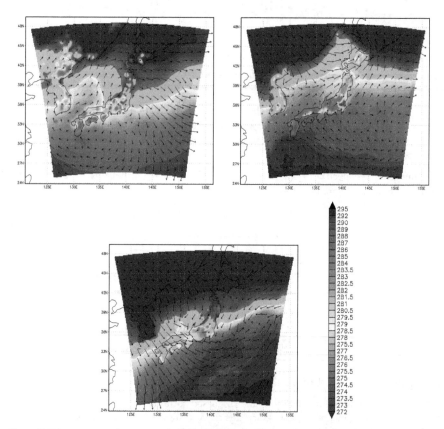

Figure 3. Temperature and mean wind profile from WRF for 3 hours, 48 hours and 93 hours after the beginning of constant release of radioactive substances.

constant release of radioactive material, the second and third plot correspond to 48 hours and 93 hours after time zero.

From the meso-scale meteorological data one may determine the eddy diffusivity coefficients for each specific hour. In the z-time plot of fig. 4 we report the dimensionless vertical eddy diffusivity coefficients as calculated by eq. (35) for four subsequent days. The figure shows in a simple way the spatial and time structure of this coefficient. In this context, it is important to point out that the largest values of the K_z coefficient correspond to the strongest mixing and likely the minimum level of contamination at ground level. It is evident analysing fig. 4 that the maximum values are reached during the day as a consequence of the strong diurnal convective mixing and in the range of a dimensionless height between $[0.4, 0.7]$, that is in the bulk of the convective boundary layer. During the night the mixing is reduced as a consequence of the formation of the stable boundary layer due to the inversion of the heat flux.

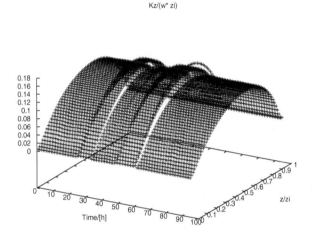

Figure 4. The dimensionless eddy diffusivity coefficient dependent on height in multiples of the boundary layer height for a time sequence of four subsequent days.

In the further we show the radioactive substance concentrations close to the surface around the nuclear power plant. Figures 5 show the distributions for 3 hours, 48 hours and 93 hours after the beginning of the substance release with a logarithmic scale.

The centre of the nuclear power plant is located in the centre of the plot, the cost line is almost in the north south direction, that is parallel to the y-axis in the plot with the ocean to the right side. Shortly after the beginning of the release the mean wind pointed towards the ocean, whereas after three days the wind blew towards the south in the direction of Tokyo.

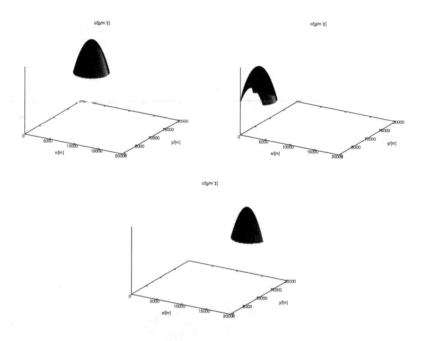

Figure 5. The logarithmic concentration distribution of radioactive substances released from the nuclear power plant for 3 hours, 48 hours and 93 hours after the beginning of the release.

7. Conclusions

The present work was based on an Eulerian approach to determine dispersion of radioactive contaminants in the PBL. To this end the diffusion equation for the cross-wind integrated concentrations was closed by the relation of the turbulent fluxes to the gradient of the mean concentration by means of eddy diffusivity (K-theory). We are completely aware of the fact that K-closure has its intrinsic limits so that one would like to remove these inconsistencies. However, comparisons of predictions by this approach to experimental data have shown that there are scenarios where this lack is not significantly manifest, which we use as a justification together with its computational simplicity to perform our simulations based on this approach.

Since the consistency of the K-approach depends crucially on the determination of the eddy diffusivity considering the turbulence structure of the PBL in its respective stability regimes, we elaborated parametrisations for the eddy diffusivity coefficients based on the micro-meteorological parameters that were extracted from meso-scale WRF simulations, that allowed to take into account the realistic orography of the larger vicinity of a reactor site in consideration. The approach proposed here for the determination of the eddy-diffusivity coefficient is based on the Taylor statistical diffusion theory and on the spectral properties of turbulence. The assumption of continuous turbulence spectrum and variances, allows the parametrisations to be continuous at all elevations, and in stability conditions ranging from a convective to a neutral condition, and from a neutral to a stable condition so that a simulation of a full diurnal cycle is possible. Simulating micro-meteorology for a short period for the Fukushima Nuclear Power Station Accident may be considered a first step into a direction where the impact of the contamination of radioactive material in the site may be simulated and evaluated for the whole period of the accident until today. Thus the present work may be understood as one tile in a larger program development that simulates radioactive material dispersion using analytical resources, i.e. solutions. In a longer term we intend to build a library that allows to predict radioactive material transport in the planetary boundary layer that extends from the micro- to the meso-scale.

The quality of the solution may be estimated by the following considerations. Recalling, that the structure of the pollutant concentration is essentially determined by the mean wind velocity $\bar{U}$ and the eddy diffusivity K, means that the quotient of norms $\omega = \frac{||K||}{||\bar{U}||}$ defines a length scale for which the pollutant concentration is almost homogeneous. Thus one may conclude that with decreasing length ($\frac{\omega}{m}$ and m an increasing integer number) variations in the solution become spurious. Upon interpreting ω^{-1} as a sampling density, one may now employ the Cardinal Theorem of Interpolation Theory ([30]) in order to find the truncation that leaves the analytical solution almost exact, i.e. introduces only functions that vary significantly in length scales beyond the mentioned limit.

The square integrable function $\chi = \int_r \bar{c} \, dt \, dx \, d\eta \in L^2$ ($\eta = y$ or z) with spectrum $\{\lambda_i\}$ which is bounded by $m\omega^{-1}$ has an exact solution for a finite expansion. This statement expresses the *Cardinal Theorem of Interpolation Theory* for our problem. Since the cut-off defines some sort of sampling density, its introduction is an approximation and is related to convergence of the approach and Parseval's theorem may be used to estimate the error. In order to keep the solution error within a prescribed error, the expansion in the region of interest has to contain $n + 1$ terms, with $n = \text{int}\left\{\frac{mL_{y,z}}{2\pi\omega} + \frac{1}{2}\right\}$. For the bounded spectrum and according to the theorem the solution is then exact. In our approximation, if m is properly chosen such that the cut-off part of the spectrum is negligible, then the found solution is almost exact.

Further, the Cauchy-Kowalewski theorem ([8]) guarantees that the proposed solution is a valid solution of the discussed problem, since this problem is a special case of the afore mentioned theorem, so that existence and uniqueness are guaranteed. It remains to justify convergence of the decomposition method. In general convergence by the decomposition method is not guaranteed, so that the solution shall be tested by an appropriate criterion. Since standard convergence criteria do not apply in a straight forward manner for the present case, we resort to a method which is based on the reasoning of Lyapunov ([6]). While Lyapunov introduced this conception in order to test the influence of variations of the initial condition on the solution, we use a similar procedure to test the stability of convergence while starting from an approximate (initial) solution R_0 (the seed of the recursive scheme). Let $|\delta Z_n| = ||\sum_{i=n+1}^{\infty} R_i||$ be the maximum deviation of the correct from the approximate solution $\Gamma_n = \sum_{i=0}^{n} R_i$, where $|| \cdot ||$ signifies the maximum norm. Then strong convergence occurs if there exists an n_0 such that the sign of λ is negative for all $n \geq n_0$. Here, $\lambda = \frac{1}{||\Gamma_n||} \log\left(\frac{|\delta Z_n|}{|\delta Z_0|}\right)$.

For model validation one faces the drawback, that the majority of measurements are at ground level, so that one could think that a two dimensional description would suffice, however the present analysis clearly shows the influence of the additional dimension. While in the two dimensional approach the tendency of the predicted concentrations is to overestimate the observed values, this is not the case for the results of the three dimensional description, mainly because it does not assume turbulence to be homogeneous. Moreover the solution of the advection diffusion equation discussed here is more general than shown in the present context, so that a wider range of applications is possible. Especially other assumptions for the velocity field and the diffusion matrix are possible. In a future work we will focus on a variety of applications and introduce a rigorous proof of convergence from a mathematical point of view, which we indicated in sketched form only in our conclusions.

Acknowledgements

The authors thank Brazilian CNPq and FAPERGS for the partial financial support of this work.

Author details

Marco Túllio Vilhena and Bardo Bodmann
Federal University of Rio Grande do Sul, Porto Alegre, RS, Brazil

Umberto Rizza
Institute ISAC, National Research Council, Lecce, Italy

Daniela Buske
Federal University of Pelotas, Pelotas, RS, Brazil

8. References

[1] Adomian, G. (1984). A New Approach to Nonlinear Partial Differential Equations. *J. Math. Anal. Appl.*, Vol. 102, page numbers (420-434).

[2] Adomian, G. (1988). A Review of the Decomposition Method in Applied Mathematics. *J. Math. Anal. Appl.*, Vol. 135, page numbers (501-544) .

[3] Adomian, G. (1994). *Solving Frontier Problems of Physics: The Decomposition Method*, Kluwer, Boston, MA.

[4] Batchelor, G.K. (1950). The application of the similarity theory of turbulence to atmospheric diffusion. *Quart. J. Royal Meteor. Soc.*, Vol. 76, No. 328, page numbers (133-146).

[5] Biagio,R., Godoy, G., Nicoli, I., Nicolli, D. & Thomas, P. (1985). *First atmospheric diffusion experiment campaign at the Angra site.* - KfK 3936, Karlsruhe, and CNEN 1201, Rio de Janeiro.

[6] Boichenko, V.A.; Leonov, G.A. & Reitmann, V. (2005). *NDimension theory for ordinary equations,* Teubner, Stuttgart.

[7] Costa, C.P.; Vilhena, M.T.; Moreira, D.M. & Tirabassi, T. (2006). Semi-analytical solution of the steady three-dimensional advection-diffusion equation in the planetary boundary layer. *Atmos. Environ.*, Vol. 40, No. 29, page numbers (5659-5669).

[8] Courant, R. & Hilbert, D. (1989). *Methods of Mathematical Physics*. John Wiley & Sons, New York.

[9] Deardoff, J.W. & Willis, G.E. (1975). A parameterization of diffusion into the mixed layer. *J. Applied Meteor.*, Vol. 14, page numbers (1451-1458).

[10] Degrazia, G.A. & Moraes, O.L.L. (1992). A model for eddy diffusivity in a stable boundary layer. *Bound. Layer Meteor.*, Vol. 58, page numbers (205-214).

[11] Degrazia, G.A.; Campos Velho, H.F. & Carvalho, J.C. (1997). Nonlocal exchange coefficients for the convective boundary layer derived from spectral properties. *Contr. Atmos. Phys.*, Vol. 70, page numbers (57-64).

[12] Demuth, C. (1978). A contribution to the analytical steady solution of the diffusion equation for line sources. *Atmos. Environ.*, Vol. 12, page numbers (1255-1258).

[13] Djolov, G.D., Yordanov, D.L. & Syrakov, D.E. (2004). Baroclinic planetary boundary layer model for neutral and stable stratification conditions. *Bound. Layer Meteor.*, Vol. 111, page numbers (467-490).

[14] Hanna, S.R. (1989). Confidence limit for air quality models as estimated by bootstrap and jacknife resampling methods. *Atmos. Environ.*, Vol. 23, page numbers (1385-1395).

[15] Højstrup, J.H. (1982). Velocity spectra in the unstable boundary layer. *J. Atmos. Sci.*, Vol. 39, page numbers (2239-2248).

[16] Irwin, J.S. (1979). A theoretical variation of the wind profile power-low exponent as a function of surface roughness and stability. *Atmos. Environ.*, Vol. 13, page numbers (191-194).

[17] Mangia, C., Degrazia, G.A. & Rizza, U. (2000). An integral formulation for the dispersion parameters in a shear/buoyancy driven planetary boundary layer for use in a Gaussian model for tall stacks. *J. Applied Meteor.*, Vol. 39, page numbers (1913-1922).

[18] Mangia, C., Moreira, D.M., Schipa, I., Degrazia, G.A., Tirabassi, T. & Rizza, U. (2002). Evaluation of a new eddy diffusivity parameterisation from turbulent eulerian spectra in different stability conditions. *Atmos. Environ.*, Vol. 36, No. 34, page numbers (67-76).

[19] Mellor, G.L. & Yamada, T. (1982). Development of a Turbulence Closure Model for Geophysical Fluid Problems. *Reviews of Geo. and Space Phys.*, Vol. 20, page numbers (851-875).

[20] Moeng, C.H. & Sullivan, P.P. (1994). A comparison of shear-and buoyancy-driven planetary boundary layer flows. *J. Atmos. Sci.*, Vol. 51, page numbers (999-1022).

[21] Moreira, D.M.; Vilhena, M.T.; Tirabassi, T.; Costa, C. & Bodmann, B. (2006). Simulation of pollutant dispersion in atmosphere by the Laplace transform: the ADMM approach. *Water, Air and Soil Pollution*, Vol. 177, page numbers (411-439).

[22] Moreira, D.M.; Vilhena, M.T.; Buske, D. & Tirabassi, T. (2009). The state-of-art of the GILTT method to simulate pollutant dispersion in the atmosphere. *Atmos. Research*, Vol. 92, page numbers (1-17).

[23] Nieuwstadt F.T.M. & de Haan B.J. (1981). An analytical solution of one-dimensional diffusion equation in a nonstationary boundary layer with an application to inversion rise fumigation. *Atmos. Environ.*, Vol. 15, page numbers (845-851).

[24] Olesen, H.R., Larsen, S.E. & Højstrup, J. (1984). Modelling velocity spectra in the lower part of the planetary boundary layer. *Bound. Layer Meteor.*, Vol. 29, page numbers (285-312).

[25] Panofsky, A.H. & Dutton, J.A. (1988). *Atmospheric Turbulence*. John Wiley & Sons, New York.

[26] Sharan, M.; Singh, M.P. & Yadav, A.K. (1996). A mathematical model for the atmospheric dispersion in low winds with eddy diffusivities as linear functions of downwind distance. *Atmos. Environ.*, Vol. 30, No.7, page numbers (1137-1145).

[27] Stroud, A.H. & Secrest, D. (1966). *Gaussian quadrature formulas*. Prentice Hall Inc., Englewood Cliffs, N.J..

[28] Taylor, G.I. (1921). Diffusion by continuous movement. *Proc. Lond. Math. Soc.*, Vol. 2, page numbers (196-211).

[29] Tirabassi T. (2003). Operational advanced air pollution modeling. *PAGEOPH*, Vol. 160, No. 1-2, page numbers (05-16).

[30] Torres, R.H. (1991). Spaces of sequences, sampling theorem, and functions of exponential type. *Studia Mathematica*, Vol. 100, No. 1, page numbers (51-74).

[31] Ulke, A.G. (2000). New turbulent parameterisation for a dispersion model in the atmospheric boundary layer. *Atmos. Environ.*, Vol. 34, page numbers (1029-1042).

[32] van Ulden, A.P. (1978). Simple estimates for vertical diffusion from sources near the ground. *Atmos. Environ.*, Vol. 12, page numbers (2125-2129).

[33] Wandel, C.F. & Kofoed-Hansen, O. (1962). On the Eulerian-Lagrangian Transform in the Statistical Theory of Turbulence. *J. Geo. Research*, Vol. 67, page numbers (3089-3093).

[34] Yordanov, D.,Syrakov, D. & Kolarova, M. (1997). On the Parameterization of the Planetary Boundary Layer of the Atmosphere: The Determination of the Mixing Height, In: *Current Progress and Problems. EURASAP Workshop Proc..*

[35] Zanetti, P. (1990). *Air Pollution Modeling.* Comp. Mech. Publications, Southampton (UK).

Permissions

The contributors of this book come from diverse backgrounds, making this book a truly international effort. This book will bring forth new frontiers with its revolutionizing research information and detailed analysis of the nascent developments around the world.

We would like to thank Wael Ahmed, for lending his expertise to make the book truly unique. He has played a crucial role in the development of this book. Without his invaluable contribution this book wouldn't have been possible. He has made vital efforts to compile up to date information on the varied aspects of this subject to make this book a valuable addition to the collection of many professionals and students.

This book was conceptualized with the vision of imparting up-to-date information and advanced data in this field. To ensure the same, a matchless editorial board was set up. Every individual on the board went through rigorous rounds of assessment to prove their worth. After which they invested a large part of their time researching and compiling the most relevant data for our readers. Conferences and sessions were held from time to time between the editorial board and the contributing authors to present the data in the most comprehensible form. The editorial team has worked tirelessly to provide valuable and valid information to help people across the globe.

Every chapter published in this book has been scrutinized by our experts. Their significance has been extensively debated. The topics covered herein carry significant findings which will fuel the growth of the discipline. They may even be implemented as practical applications or may be referred to as a beginning point for another development. Chapters in this book were first published by InTech; hereby published with permission under the Creative Commons Attribution License or equivalent.

The editorial board has been involved in producing this book since its inception. They have spent rigorous hours researching and exploring the diverse topics which have resulted in the successful publishing of this book. They have passed on their knowledge of decades through this book. To expedite this challenging task, the publisher supported the team at every step. A small team of assistant editors was also appointed to further simplify the editing procedure and attain best results for the readers.

Our editorial team has been hand-picked from every corner of the world. Their multi-ethnicity adds dynamic inputs to the discussions which result in innovative outcomes. These outcomes are then further discussed with the researchers and contributors who give their valuable feedback and opinion regarding the same. The feedback is then collaborated with the researches and they are edited in a comprehensive manner to aid the understanding of the subject.

Apart from the editorial board, the designing team has also invested a significant amount of their time in understanding the subject and creating the most relevant covers. They scrutinized every image to scout for the most suitable representation of the subject and create an appropriate cover for the book.

The publishing team has been involved in this book since its early stages. They were actively engaged in every process, be it collecting the data, connecting with the contributors or procuring relevant information. The team has been an ardent support to the editorial, designing and production team. Their endless efforts to recruit the best for this project, has resulted in the accomplishment of this book. They are a veteran in the field of academics and their pool of knowledge is as vast as their experience in printing. Their expertise and guidance has proved useful at every step. Their uncompromising quality standards have made this book an exceptional effort. Their encouragement from time to time has been an inspiration for everyone.

The publisher and the editorial board hope that this book will prove to be a valuable piece of knowledge for researchers, students, practitioners and scholars across the globe.

List of Contributors

Chang-Hsing Lee
EE Dep. of National Tsing Hua University, Hsinchu, Taiwan

Shi-Lin Chen
EE Dep. of Chung Yuan Christian University, Chung Li, Taiwan

Wael H. Ahmed
Department of Mechanical Engineering, King Fahd University of Petroleum & Minerals, Dhahran, Saudi Arabia

Luciano Burgazzi
Reactor Safety and Fuel Cycle Methods Technical Unit, ENEA, Italian National Agency for New Technologies, Energy and Sustainable Economic Development, Bologna, Italy

Prabhakar Sharma
Department of Earth Sciences, Uppsala University, Uppsala, Sweden

Tamás János Katona
Nuclear Power Plant Paks Ltd., Hungary

Heinz Peter Berg
Federal Office for Radiation Protection (BfS), Department of Nuclear Safety, Salzgitter, Germany

Jan Hauschild
TÜV NORD SysTec GmbH & Co. KG, Hamburg, Germany

Vladimir M. Kotov
Department of Development and Test of Reactor Devices, Institute of Atomic Energy of NNC RK, Kurchatov, Kazakhstan

Marco Túllio Vilhena and Bardo Bodmann
Federal University of Rio Grande do Sul, Porto Alegre, RS, Brazil

Umberto Rizza
Institute ISAC, National Research Council, Lecce, Italy

Daniela Buske
Federal University of Pelotas, Pelotas, RS, Brazil